T0325126

An Analysis of

Brundtland Commission's

Our Common Future

Ksenia Gerasimova

Published by Macat International Ltd
24:13 Coda Centre, 189 Munster Road, London SW6 6AW.

Distributed exclusively by Routledge
2 Park Square, Milton Park, Abingdon, Oxon OX14 4RN
711 Third Avenue, New York, NY 10017, USA

Routledge is an imprint of the Taylor & Francis Group, an informa business

www.macat.com
info@macat.com

Cataloguing in Publication Data
A catalogue record for this book is available from the British Library.
Library of Congress Cataloguing-in-Publication Data is available upon request.
Cover illustration: Etienne Gilfillan

ISBN 978-1-912302-34-5 (hardback)
ISBN 978-1-912128-75-4 (paperback)
ISBN 978-1-912281-22-0 (e-book)

Notice
The information in this book is designed to orientate readers of the work under analysis,
to elucidate and contextualise its key ideas and themes, and to aid in the development
of critical thinking skills. It is not meant to be used, nor should it be used, as a
substitute for original thinking or in place of original writing or research. References and
notes are provided for informational purposes and their presence does not constitute
endorsement of the information or opinions therein. This book is presented solely for
educational purposes. It is sold on the understanding that the publisher is not engaged
to provide any scholarly advice. The publisher has made every effort to ensure that
this book is accurate and up-to-date, but makes no warranties or representations with
regard to the completeness or reliability of the information it contains. The information
and the opinions provided herein are not guaranteed or warranted to produce particular
results and may not be suitable for students of every ability. The publisher shall not be
liable for any loss, damage or disruption arising from any errors or omissions, or from
the use of this book, including, but not limited to, special, incidental, consequential or
other damages caused, or alleged to have been caused, directly or indirectly, by the
information contained within.

CONTENTS

WAYS IN TO THE TEXT

Who Is Gro Brundtland?   9

What Does *Our Common Future* Say?   10

Why Does *Our Common Future* Matter?   12

SECTION 1: INFLUENCES

Module 1: The Author and the Historical Context   16

Module 2: Academic Context   20

Module 3: The Problem   24

Module 4: The Author's Contribution   29

SECTION 2: IDEAS

Module 5: Main Ideas   34

Module 6: Secondary Ideas   38

Module 7: Achievement   43

Module 8: Place in the Author's Work   48

SECTION 3: IMPACT

Module 9: The First Responses   54

Module 10: The Evolving Debate   59

Module 11: Impact and Influence Today   64

Module 12: Where Next?   69

Glossary of Terms   74

People Mentioned in the Text   81

Works Cited   86

THE MACAT LIBRARY

The Macat Library is a series of unique academic explorations of seminal works in the humanities and social sciences – books and papers that have had a significant and widely recognised impact on their disciplines. It has been created to serve as much more than just a summary of what lies between the covers of a great book. It illuminates and explores the influences on, ideas of, and impact of that book. Our goal is to offer a learning resource that encourages critical thinking and fosters a better, deeper understanding of important ideas.

Each publication is divided into three Sections: Influences, Ideas, and Impact. Each Section has four Modules. These explore every important facet of the work, and the responses to it.

This Section-Module structure makes a Macat Library book easy to use, but it has another important feature. Because each Macat book is written to the same format, it is possible (and encouraged!) to cross-reference multiple Macat books along the same lines of inquiry or research. This allows the reader to open up interesting interdisciplinary pathways.

To further aid your reading, lists of glossary terms and people mentioned are included at the end of this book (these are indicated by an asterisk [*] throughout) – as well as a list of works cited.

Macat has worked with the University of Cambridge to identify the elements of critical thinking and understand the ways in which six different skills combine to enable effective thinking.
Three allow us to fully understand a problem; three more give us the tools to solve it. Together, these six skills make up the **PACIER** model of critical thinking. They are:

ANALYSIS – understanding how an argument is built
EVALUATION – exploring the strengths and weaknesses of an argument
INTERPRETATION – understanding issues of meaning

CREATIVE THINKING – coming up with new ideas and fresh connections
PROBLEM-SOLVING – producing strong solutions
REASONING – creating strong arguments

To find out more, visit **WWW.MACAT.COM.**

CRITICAL THINKING AND *OUR COMMON FUTURE*

Primary critical thinking skill: ROBLEM-SOLVING
Secondary critical thinking skill: EVALUATION

Our Common Future is a joint work produced in 1987 by a United Nations commission headed by former Norwegian Prime Minister, Gro Brundtland. Also known as *The Brundtland Report,* it is a striking example of the critical thinking skill of asking productive questions in order to produce valid solutions to a problem.

Brundtland's report asks how we can protect the world we live in for future generations, while at the same time stimulating economic and social development now. The solution the work proposes is 'sustainable development, defined as humanity's ability 'to ensure that it meets the needs of the present without compromising the ability of future generations to meet their own needs.' The key conclusion the report came to – that we need long-term strategies to manage the earth's natural resources – proved to be so universally welcomed it introduced the term 'sustainability' into the everyday language of international politics.

Solving the problem of workable sustainable development became a hot topic, leading to the birth of a new academic discipline, environmental economics. By asking the right questions – and critically evaluating the answers to those questions – *Our Common Future* offered a solution to the problem of ensuring sustainable development by highlighting the critical importance of international cooperation.

ABOUT THE AUTHOR OF THE ORIGINAL WORK

Gro Harlem Brundtland was born in Oslo, Norway, in 1939, and trained as a doctor before becoming a politician. She served as minister for environmental affairs from 1974 to 1979 and then, in 1981, became prime minister of Norway—the first woman to do so. In 1983 she was recruited to chair the United Nations World Commission on Environment and Development and to oversee its report, *Our Common Future*, which was published in 1987. Brundtland served two more terms as prime minister, from 1986 to 1989 and from 1990 to 1996. Today, she is a prominent political figure and deputy chair of the Elders, an influential group of world leaders.

ABOUT THE AUTHOR OF THE ANALYSIS

Dr Ksenia Gerasimova holds a PhD from the University of Cambridge, where she is currently an Affiliated Lecturer at the Centre for Development Studies.

ABOUT MACAT

GREAT WORKS FOR CRITICAL THINKING

Macat is focused on making the ideas of the world's great thinkers accessible and comprehensible to everybody, everywhere, in ways that promote the development of enhanced critical thinking skills.

It works with leading academics from the world's top universities to produce new analyses that focus on the ideas and the impact of the most influential works ever written across a wide variety of academic disciplines. Each of the works that sit at the heart of its growing library is an enduring example of great thinking. But by setting them in context – and looking at the influences that shaped their authors, as well as the responses they provoked – Macat encourages readers to look at these classics and game-changers with fresh eyes. Readers learn to think, engage and challenge their ideas, rather than simply accepting them.

WAYS IN TO THE TEXT

KEY POINTS

- Born in 1939, Gro Brundtland was the first female prime minster of Norway and is a prominent figure in the United Nations* (or UN, an international organization founded to foster cooperation between governments); in 1983 she was appointed as the chair of the UN World Commission on Environment and Development* (the Brundtland Commission).

- As a trained medical doctor and an experienced politician, Brundtland came up with a visionary approach to managing environmental and social resources, by understanding that environmental challenges and human development* (political and social processes aimed at increasing people's life choices) were fundamentally linked.

- *Our Common Future* (1987), also known as the *Brundtland Report*, is one of the main reference works on sustainable development* (the means by which societies can support themselves while preserving natural resources); it has provided the most cited definition of "sustainability"* in both academic and policymaking* circles.

Who Is Gro Brundtland?

Gro Harlem Brundtland, the author of *Our Common Future* (1987), was born in 1939 in Oslo, Norway. Her father, Gudmund Harlem, was a medical doctor and a prominent politician, like Brundtland. Speaking of her father's influence on her career choice, she says, "I'm proud of my father for voting Labor [the Norwegian Labor Party*: a social-democratic political party] because it's right, even though he does not benefit from it."[1] Brundtland studied medicine at the Harvard School of Public Health in the United States, where she was exposed to "the seeds of internationalism."[2] ("Internationalism" here refers to a political stance that favors cooperation between different states and peoples.)

Brundtland began work as a doctor, but in 1974 she put aside her medical career to become Norway's minister for environmental affairs. In 1981, she became prime minister of Norway, the first woman to hold this position. Brundtland recognized the connection between human health and the state of the environment.[3] Her background was unique in that she was the only prime minister who had ever held an environmental post. The then-secretary-general of the UN, Javier Perez de Cuellar,* appointed her to chair the World Commission on Environment and Development and to lead a "call for political action" on environment and development.[4]

Brundtland gathered an extraordinary group of commissioners, both academics and politicians. Together they came up with visionary ideas about human development and compiled their findings in a report under the title *Our Common Future*. This has since been referred to as "the most important document of the decade on the future of the world," according to Oxford University Press.[5]

What Does *Our Common Future* Say?

Our Common Future was a report commissioned to formulate "a global agenda for change"[6] and produced by the Brundtland Commission,* a body convened by the UN with the aim of fostering cooperation between nations to pursue sustainable development. It was conceived as a response to a growing number of serious concerns faced by the global community in the 1980s. The report recognized the need for long-term strategies for the management of natural resources. The Brundtland commissioners agreed that it was necessary to protect the environment, while at the same time stimulating economic and social development through international cooperation. Taken together, these goals were summarized as "sustainable development,"* defined in the report as humanity's ability "to ensure that it meets the needs of the present without compromising the ability of future generations to meet their own needs."[7] This recognition of our responsibility to the future is captured in the title of the report.

Our Common Future has been translated into over 20 languages and was first distributed to politicians, together with accompanying educational video materials.[8] The report is an appeal to humanity to change its behavior in order to reverse dangerous environmental and social trends, both on an individual and a global level. In 1987, it was published as a book by Oxford University Press, a unique occurrence for a work that started life as a UN report.[9] By 1989, half a million copies had been sold.[10]

Our Common Future introduced the term "sustainability" into the everyday language of international politics. In the world of academia, discussion of sustainable development has led to the birth of a new discipline of environmental economics* (the study of economic matters as they are affected by issues of sustainability, and of the economic consequences of policies designed to protect the environment) and has also received attention from many other disciplines, such as human geography (the study of interactions between human populations and the environment), international relations (the study of the relations between and among nations and international institutions), and social anthropology (the study of the ways in which humans build and maintain social institutions such as laws and kinship). *Our Common Future* still serves as a reference book for both academic research and policymaking.

It is hard to overstate the importance and influence of *Our Common Future*. The points made in the report have become generally accepted, and Gro Brundtland's expert advice has been sought by a number of UN agencies, including the General Assembly* (the body of the UN in which all the world's nations are represented) and the World Health Organization* (or WHO, the agency dealing with global public health). The current secretary-general of the UN, Ban Ki-moon,* has appointed Brundtland one of his three special envoys* on environmental issues and climate change.* ("Special envoy" here refers to a person chosen by the secretary-general to act as a representative and adviser on a particular matter.)

Why Does *Our Common Future* Matter?

Our Common Future is widely acknowledged as the main source for the concept of sustainable development. It has introduced the term "sustainability" and the conceptual cornerstones of contemporary development policies—the three pillars of economic growth, social development (an improvement in matters such as living standards, healthcare, and institutions like the courts) and conservation of nature. While Brundtland and her commissioners were not the first to use this term, they were the first to give it such a wide international exposure and, as a result, they have made it a frequently used concept.

The report raised universal issues related to balancing human requirements with responsible management of the planet's natural resources. The book's insights into the need to stop the harmful overuse of natural resources were a revelation to many readers.[11] The authors were among the first in the field of international policy to connect environmental protection and economic development. Seeing the fulfillment of human needs as the major goal of development,[12] they mapped out the serious negative effects of damage to the environment, explaining how these threatened human well-being in the present and the future.

Brundtland and her commissioners cataloged and analyzed the major challenges faced by human societies at the time of publication; among these were poverty, deterioration of finite natural resources, and climate change. They then discussed possible solutions on a global and national scale, such as better international cooperation and raising awareness about the environment among politicians and the general public. The writers recognized the global scale of the challenges discussed in the report, believing that changes in international policy and increased awareness were the key to sustainability.

Having established the main conceptual framework of sustainability, the commissioners discussed the future and made what seems in retrospect an overly optimistic estimation that humanity would complete the global shift into sustainability by the year 2000. The world community still plans to reach Sustainable Development Goals (SDGs)* by 2030, though this is a full 30 years later than the original schedule proposed by Brundtland.

The Brundtland commissioners also identified the organizations and people who might bring about this change. They predicted that national governments would be resistant to the radical changes they proposed. As a result, the main push for policy change was expected to come from civil society* groups—political groups made up of ordinary people rather than professional politicians—who would work with both the general public and policymakers.

NOTES

1   Gro Harlem Brundtland, *Madam Prime Minister: A Life in Power and Politics* (New York: Farrar, Straus and Giroux, 2005), 19.

2   United Nations, *Biography of Dr Gro Harlem Brundtland* (Geneva: United Nations, 2014), 1.

3   World Health Organization, "Dr Gro Harlem Brundtland, Director-General." WHO: Geneva, 1998, accessed March 9, 2016, http://www.who.int/dg/brundtland/bruntland/en/.

4   World Commission on Environment and Development (WCED), *Our Common Future* (Oxford: Oxford University Press, 1989), X.

5   WCED, *Our Common Future*, publisher's note, cover page.

6   WCED, *Our Common Future*, viii.

7   WCED, *Our Common Future*, 8.

8   Pratap Chatterjee and Matthias Finger, *The Earth Brokers: Power, Politics and World Development* (London: Routledge, 2013), 84.

9   WCED, Report of the World Commission and Development on Environment and Development: *Our Common Future*, August 4, 1987, accessed February 1, 2016, http://www.un-documents.net/wced-ocf.htm.

10   Linda Starke, *Signs of Hope: Working towards Our Common Future* (Oxford: Oxford University Press, 1990), 3.

11   Winin Pereira and Jeremy Seabrook, *Asking the Earth: Farms, Forests and Survival in India* (Sterling, VA: Earthscan, 1990), 62.

12   WCED, *Our Common Future*, 44.

SECTION 1
INFLUENCES

MODULE 1
THE AUTHOR AND THE
HISTORICAL CONTEXT

KEY POINTS

- *Our Common Future* (the *Brundtland Report*) provided the classic definition of sustainable development* still used today.
- The report inspired an important shift in international political thinking, raising awareness about the urgent need to manage natural resources in a responsible manner so that they would be available for future generations.
- Major challenges named in the report, such as climate change,* are still relevant today.

Why Read This Text?

Gro Brundtland's *Our Common Future* (1987) is the key text for the concept of "sustainable development": human society's ability to ensure that the various economic and institutional improvements necessary to increase a population's life choices "meets the needs of the present without compromising the ability of future generations to meet their own needs."[1] The text's publication marked the beginning of an important period in our history, as it was the first time that high-level politicians recognized that the business-as-usual approach to managing the world's natural resources would eventually lead to global environmental collapse. The authors called for a total rethink of production and distribution at a global level.

The report gave the first comprehensive outline of major issues such as a fast-growing world population, food insecurity* (lack of access to food), increasing urban populations, environmental

> **❝** I decided to accept the challenge. The challenge of facing the future, and of safeguarding the interests of coming generations. For it was abundantly clear: We needed a mandate for change. **❞**
>
> Gro Brundtland, Chairman's Foreword, *Our Common Future*

degradation, and climate change. By March 1989, 22 national governments had created plans for achieving sustainability,* demonstrating their acceptance of the report and of the new concept of sustainable development.[2] The report continues to influence policymaking* processes today.

Author's Life

Gro Brundtland was born in 1939 in the Norwegian capital, Oslo. She was trained as a medical doctor, following in the footsteps of her father, who was a doctor and a prominent Norwegian politician. She went on to study at the Harvard School of Public Health, where she refined her views on health and human development.* From an early age she had been involved in political activism, her father having enrolled her as a member of the left-wing Norwegian Labor* Movement when she was just eight years old.

On returning to Norway in 1965 after completing her graduate studies at Harvard, Brundtland joined the Norwegian Ministry of Health, where she worked as a medical consultant. In 1974, she accepted a nomination to become minister for environmental affairs and in 1981 she became the youngest person and the first woman to serve as prime minister of Norway. She stayed in office for two terms (1986–9 and 1990–6). Her unique background as a minister for the environment and a prime minister later led to her being chosen to lead the World Commission on Environment and Development*—a body instituted by the United Nations* to find solutions to the pressing problem of development in a world of finite resources.

Brundtland assembled an international team of leaders to author *Our Common Future*, consisting of 21 prominent academics, politicians, and officials from different nations. Brundtland described the work of her commission as in "the spirit of friendship and open communication," reflecting the learning process of people with "different views and perspectives, different values and beliefs, and very different experiences and insights." Despite this diversity, the report's writers arrived at a shared conclusion about the urgent need for sustainability.[3]

Author's Background

Gro Brundtland followed her father in becoming a prominent Norwegian politician and a practicing medical doctor. Before she was nominated to lead the writing of *Our Common Future*, she had participated in other UN commissions that dealt with questions of disarmament (the decommissioning of weapons of war), poverty, and the environment. As an experienced and internationally recognized politician, she was familiar with the main issues of the time.

Our Common Future can be considered a product of its era. It reflects the urgency of the task of formulating sustainable development strategies in "a compelling reality" of political, economic and environmental problems.[4] The Cold War* (1947–91), a period of grave tension between the United States and its allies and the now-dissolved Soviet Union* and its allies, was still ongoing at the time. The world community faced a proliferation of massive public (that is, state) debt in developing countries, reduced aid payments from richer to poorer countries, and growing poverty.[5] Famines in Africa, the Indian Bhopal* disaster of 1984 (the worst industrial disaster in history) and the Russian Chernobyl* accident of 1986 (the worst nuclear power-plant disaster to date) are mentioned in the report as examples of how the future of humanity could be endangered unless changes in attitudes and policies were made.[6]

Many of the challenges mentioned in the report still remain important today, such as climate change, food insecurity, and environmental degradation. This explains why Gro Brundtland is still actively involved in international politics as a special envoy to the current UN secretary-general, Ban Ki-moon* of Korea, on climate change.

NOTES

1 World Commission on Environment and Development (WCED), *Our Common Future* (Oxford: Oxford University Press, 1989), 8.

2 Linda Starke, *Signs of Hope: Working towards Our Common Future* (Oxford: Oxford University Press, 1990), 3.

3 WCED, *Our Common Future*, xiii.

4 WCED, *Our Common Future*, xix.

5 WCED, *Our Common Future*, x.

6 WCED, *Our Common Future*, 3.

MODULE 2
ACADEMIC CONTEXT

KEY POINTS

- *Our Common Future* raised the important question of how human society can prosper despite decreasing natural resources.

- Disagreeing with the assumption that a growing human population would have catastrophic consequences, the members of the Brundtland Commission* remained optimistic that a new concept of sustainability* — the requirement that we live within our finite natural resources — could address many of these concerns.

- The report presented the hopes and ideals of its authors and their contemporaries about a better future for humanity.

The Work in its Context

Gro Brundtland's report *Our Common Future* (1987) has been recognized for its contribution to the evolution of the concept of sustainable development—the capacity to increase living standards while ensuring that future generations will not find themselves without resources in an environmentally degraded world. The report recognized the concepts of "development"* and "environment" as inseparable, since it was becoming clear that failures of development policies and mismanagement of natural resources were deeply interconnected.[1]

The concept of sustainability first came to public notice in the 1970s, in the American environmentalist Wes Jackson's* work on agriculture. Jackson argued that nonrenewable resources should be used in the most efficient possible way.[2]

> ❝ The challenge of finding sustainable development paths ought to provide the impetus—indeed the imperative—for a renewed search for multilateral solutions and a restructured international economic system of co-operation. These challenges cut across the divides of national sovereignty, of limited strategies for economic gain, and of separated disciplines of science. ❞
>
> World Commission on Environment and Development,* *Our Common Future*

The question of how humans can survive even though the planet's available resources are finite was discussed before *Our Common Future*. For example, in 1972 the Club of Rome* (an international think tank* that united academics, businessmen, and politicians) used computer modeling to produce the influential report *The Limits to Growth*.[3] The report offered statistical evidence identifying future trends of overpopulation and growing demand for natural resources; the authors warned that these would be irreversible unless a better, more sustainable, approach to the management of natural resources was put in place.

Overview of the Field

The relationship between finite environmental resources and human survival has been widely discussed since the work of the eighteenth-century English demographer Thomas Malthus* ("demography" refers to the statistical study of the makeup of a community or population). His *An Essay on the Principle of Population* (1798) argued that there is a link between population growth, food availability, and a limited "power in the earth to produce subsistence for man."[4] Malthus introduced a hypothetical scenario in which a world population could outreach the food supply available, and argued that population growth should be limited—similar issues to those identified in the Club of Rome's report.

In the period just before the publication of *Our Common Future* it was already becoming clear that the earth's resources had to be managed sustainably. This idea challenged the theory, then dominant in mainstream economics, that the problems presented by finite environmental resources and environmental degradation could be solved by the relocation of people from one region of the world to another.[5]

Academic Influences

There were earlier attempts to develop the concept of ecological sustainability at the United Nations* Conference on the Human Environment held in Sweden in 1972, also known as the Stockholm Conference.* This was the first global conference to evaluate the environmental consequences of human development at a macro (that is, general and global) level and to elaborate universal guidelines to protect and improve the environment.[6] The conference passed a Declaration on Human Development, which acknowledged that every individual has a right to a healthy environment. National governments were given responsibility for implementing and safeguarding policies to protect the natural environment and the UN officially recognized the importance of addressing the world's environmental problems.[7]

In the same year, the Club of Rome issued its report *The Limits to Growth*, identifying future trends such as overpopulation and a growing demand for natural resources. The report was well received, and many politicians were immediately convinced by its argument about the correlation between overpopulation and the devastation of natural resources. Later on, however, the Club of Rome was criticized for using incorrect data.[8] One of the Brundtland commissioners, Japanese economist Saburo Okita,* was also a member of the Club of Rome.

The Brundtland commissioners were aware of the ongoing academic and political debates around global environmental problems. They were the experts in the field at the time, and called for more international cooperation. The commissioners argued, for example, that countries of the Global North* (richer countries, mostly in the northern hemisphere) should offer more consistent support to their poorer counterparts in the Global South* (impoverished countries, mostly in the southern hemisphere).

NOTES

1   World Commission on Environment and Development, *Our Common Future* (Oxford: Oxford University Press, 1989), 30.

2   Wes Jackson, *New Roots for Agriculture* (Lincoln: University of Nebraska Press, 1985), 144.

3   See Donella H. Meadows et al., *The Limits to Growth: A Report for the Club of Rome's Project on the Predicament of Mankind* (New York: Universe Books, 1974).

4   Thomas Robert Malthus, *An Essay on the Principle of Population* (London: J. Johnson, 1798), 13.

5   Jorgen Norgard et al., "The History of the Limits to Growth," *Solutions* 2, no.1 (2010): 59–63.

6   Felix Dodds et al., *Only One Earth: The Long Road via Rio to Sustainable Development* (London: Routledge, 2012), 8–11.

7   Dodds et al., *Only One Earth*, 11–12.

8   Graham Turner, *A Comparison of Limits to Growth with Thirty Years of Reality*, CSIRO Working Paper (Canberra: CSIRO, 2008), 36.

MODULE 3
THE PROBLEM

KEY POINTS

- The core question of *Our Common Future* was how to ensure the economic growth necessary for social development* while using and managing finite natural resources in a responsible way.

- The topics addressed in the report had previously been addressed by other contemporary scholars, including the think tank* the Club of Rome,* the American environmentalist Lester Brown,* and those engaged in the new field of environmental economics.*

- The authors of the report were criticized for prioritizing economic growth over the protection of natural resources; only one of them directly responded to this criticism.

Core Question

In *Our Common Future*, Gro Brundtland addressed four strategic goals:
- "To propose long-term environmental strategies for achieving sustainable development* by the year 2000."
- "To achieve greater cooperation among developed and developing countries."
- "To optimize strategies for addressing environment concerns."
- "To help define shared perceptions of long-term environmental issues."[1]

The core question of the report was how to secure the survival and well-being of humankind in the future through safeguarding the environment.[2] The Brundtland commissioners* approached this question through an understanding that the planet's existing ecological

> **❝** The environment does not exist as a sphere separate
> from human actions, ambitions, and needs, and attempts
> to defend it in isolation from human concerns have
> given the very word 'environment' a connotation of
> naivety in some political circles. **❞**
> World Commission on Environment and Development,* *Our Common Future*

system could not support the growing needs of humans without a major shift in international policies related to natural-resource management. The imbalance between limited environmental resources and growing human needs was illustrated by problems such as economic crises, deepening global poverty, and life-threatening accidents, both natural and man-made. For example, major industrial accidents in India at a pesticide factory in Bhopa* and in what is today Ukraine at the Chernobyl* nuclear power-plant, both during the 1980s, led to deaths, injuries, and long-term negative environmental effects. These events forced the global community to reassess the danger posed by technogenic catastrophes (accidents caused by technology). Better international cooperation was discussed as a way to prevent this kind of incident.[3]

While no early drafts of the report have been published, Brundtland commissioners have revealed that the initial title of the report was *A Threatened Future*.[4] The commissioners were dedicated to defending humanity against future environmental and social crises.

The Participants

An international team, known as a "commission," wrote the report under Brundtland's leadership. It was composed of 21 prominent academics, politicians, and international officials from Algeria, Brazil, Canada, China, Colombia, Cote d'Ivoire, Germany, Guyana, Hungary,

India, Indonesia, Italy, Japan, Nigeria, Norway, Saudi Arabia, the Soviet Union,* Sudan, the United States, Yugoslavia,* and Zimbabwe. Among them were Maurice Strong,* a Canadian businessman and the first president of the United Nations Environment Program (UNEP),* Nitin Desai,* an Indian economist, and Saburo Okita,* a former Japanese minister and member of the Club of Rome, who led Japan to its postwar economic growth. The 21 participants who produced the report are referred to as "commissioners."

The commission was an independent international body and aimed to provide a full analysis based on numerous qualitative and quantitative data sets—a mixture of numerical information and other kinds of research, such as descriptions and interviews. To ensure that the most up-to-date information was used, the commission held five public hearings with indigenous people, farmers, and scientists. The authors described *Our Common Future* as written collaboratively by many people in "all walks of life."[5]

There had been previous attempts to provide a comprehensive understanding of how to manage earth's diminishing resources, such as *The Limits to Growth* by the Club of Rome, and articles and books by the American environmentalist Lester Brown.[6] Both the Club of Rome and Brown considered a number of possible scenarios for managing population growth and finite natural resources, sometimes even predicting apocalyptic outcomes in which competition for access to natural resources would bring about a final global catastrophe. In contrast, the Brundtland commissioners had faith in humanity's ability to solve these problems peacefully by means of sustainable development. With the exception of one commissioner, the Indian economist Nitin Desai, who eventually rejected the concept, every member of the team that produced *Our Common Future* continued to use the concept of sustainable development in his or her later work.

The Contemporary Debate

Although the report introduced the concept of sustainable development as we know it today, it did not explain how sustainability could be achieved through the management of natural resources. It adopted an anthropocentric approach (one centering on human needs), rather than a nature-centric, or ecocentric, approach (one emphasizing nature above human needs).[7] Overall, *Our Common Future* supports an economic argument known as "weak* sustainability," an approach to the management of natural capital* (natural resources beneficial to human prosperity) that sees it as potentially interchangeable with or replaceable by human capital* (knowledge, technology, and innovation).

For this reason, the report has been heavily criticized by ecological economists, who argue that it falsely offers "a sense of comfort to the effect that the environment can be dispensed with," [8] and implies incorrectly that implementing sustainability will not require much radical change. Brundtland commissioners addressed this critique in different ways. One member of the team, Nitin Desai, changed his views about sustainability in response to criticisms. He later voiced opposition to some of the content of the report he had helped to produce.[9]

In contrast, another Brundtland commissioner, Canadian businessman and environmentalist Maurice Strong, remained committed to the views expressed in the report. Strong understood sustainable development as a set of policies arranged "on the premise that population and per capita consumption [consumption per head of population] must operate within the global ecosystem* to respond indefinitely to our demand on resources and to assimilate the wastes produced."[10] (An "ecosystem" is a biological system comprising all the organisms that exist in a specific physical environment.)

NOTES

1   World Commission on Environment and Development (WCED), *Our Common Future* (Oxford: Oxford University Press, 1989), VIII

2   WCED, *Our Common Future*, xi.

3   WCED, *Our Common Future*, 95–235.

4   Linda Starke, *Signs of Hope: Working towards Our Common Future* (Oxford: Oxford University Press, 1990), 1.

5   WCED, *Our Common Future*, xiii.

6   Lester R. Brown, *Man, Land and Food: Looking Ahead at World Food Needs* (Washington, DC: US Department of Agriculture, 1963).

7   Parenivel Mauree, "Sustainability and Sustainable Development," *Le Mauricien*, August 17, 2011, accessed February 1, 2016, http://www.lemauricien.com/article/maurice-ile-durable-sustainability-and-sustainable-development.

8   P. A. Victor et al., "How Strong is Weak Sustainability?" *Economie Appliquée* 48, no. 2 (1995): 75–94.

9   Nitin Desai, "Symposium: the Road from Johannesburg," keynote address (Georgetown: *Environmental Law Review*, 2003).

10 Maurice Strong, *Where on Earth We are Going?* (Toronto: Vintage Canada, 2001), 195.

MODULE 4
THE AUTHOR'S CONTRIBUTION

KEY POINTS

- Gro Brundtland was determined to articulate the way in which the environment and development* are interconnected, and to emphasize the need to manage natural resources in a responsible manner.

- At the time of publication of *Our Common Future*, the idea of sustainability* was an almost revolutionary way of thinking about managing economies in that it took natural resources into account.

- Brundtland's previous involvement in international commissions gave her extensive knowledge of progressive ideas about international development.

Author's Aims

Gro Brundtland's work on *Our Common Future* began in December 1983 when the secretary-general of the United Nations,* Javier Perez de Cuellar,* appointed her chair of the World Commission on Environment and Development. Then the Norwegian prime minister, she saw this appointment as "the challenge of facing the future, and of safeguarding the interests of coming generations"[1] and as an opportunity to work toward the ideals she had developed as a participant in two previous UN commissions, the *Brandt Report** on global economic development and the Palme Commission* on disarmament and global security. In her role as the chair of the Palme Commission, she aimed to achieve two main interconnected goals: to persuade political leaders to return to multilateral* disarmament (that is, a policy implemented simultaneously by a number of countries

> 66 But the 'environment' is where we all live; and 'development' is what we all do in attempting to improve our lot within that abode. The two are inseparable. Further, development issues must be seen as crucial by the political leaders who feel that their countries have reached a plateau towards which other nations must strive. 99
>
> World Commission on Environment and Development,* *Our Common Future*

working together), despite the threat of nuclear war that characterized the ongoing Cold War* between the United States and the Soviet Union* and their allies; and to expand the understanding of human development.

Although disappointed by the deterioration in global cooperation throughout the 1970s and 1980s, Brundtland remained hopeful that progressive ideas could be brought into reality as a result of a new awareness promoted at the international level. By raising concerns about instances of environmental degradation, she aimed to prove that these were shared human challenges that required a unified international response and a more equitable distribution of global financial resources. In *Our Common Future*, environmental concerns were presented as closely related to social concerns. For example, climate change* and soil depletion (infertility in soil arising from overuse by farmers) were recognized as complex problems threatening human survival that could not be solved by developing (poorer) nations on their own, but required aid from richer nations.

Approach
Environmental concerns had been discussed at previous UN conferences such as the Stockholm Conference* of 1972. These conferences provided precedents for the Brundtland Commission.

The most important achievement of the commission was the recognition of the link between the state of the natural environment and human well-being. Brundtland aimed to develop these ideas further in *Our Common Future*. She challenged the initial terms of reference for her commission, insisting that its considerations should not be limited to environmental issues; instead, she argued that environmental concerns should be understood in relation to human development. *Our Common Future* stated that the environment and development were in fact inseparable. The major concept of the report—sustainable development*—was formulated on the basis of this idea.

According to the report, sustainable development has limits created by the current level of technological development, existing social organizations, and the planet's ability to renew its resources. The report also introduced the element of economic growth into the understanding of development. Sustainable development was presented as a three-part challenge comprising environmental protection, social development (the bettering of human life choices), and economic growth.

Contribution in Context

Our Common Future's expansion of the meaning of environmental concerns was novel. Brundtland and her colleagues argued that the environment should not be considered "as a sphere separate from human actions, ambitions, and needs." They added that "attempts to defend it in isolation from human concerns have given the very word 'environment' a connotation of naivety in some political circles."[2] For the first time, the UN took the official position that the environment was interconnected with human development and economic growth.

The report's mention of the needs of future generations was also new. Unlike most UN reports, designed as internal documents to be circulated only within UN agencies, *Our Common Future* was published as a book. This was a result of its novelty and the urgency of the shift toward multilateral international development suggested in the report. The report called on all the world's citizens to work together to bring about the sustainability required to ensure human survival.

NOTES

1   World Commission on Environment and Development (WCED), *Our Common Future* (Oxford: Oxford University Press, 1989), ix.

2   WCED, *Our Common Future*, x.

SECTION 2
IDEAS

MODULE 5
MAIN IDEAS

KEY POINTS

- *Our Common Future* identified major challenges faced by humanity in the twentieth century, including environmental crises, poverty, the depletion of natural resources, climate change,* and political tensions such as the Cold War.*

- The report defines sustainability* as humanity's ability "to ensure that it meets the needs of the present without compromising the ability of future generations to meet their own needs."

- The report aimed to raise awareness among politicians and the general public about the need to change the pattern of overconsumption of natural resources.

Key Themes

The core question posed by Gro Brundtland's *Our Common Future* (1987) was how to secure the survival and well-being of humankind in the distant future while safeguarding the environment.[1] The Cuban missile crisis* of 1962—perhaps the closest the United States came to nuclear war with its Cold War adversary, the Soviet Union*—and a number of industrial accidents with serious environmental effects, such as the Indian Bhopal* disaster and the Soviet nuclear power-plant catastrophe at Chernobyl,* all had long-term negative effects on the natural environment and human health. Politicians and other Western leaders, including Brundtland and her commissioners, wanted to prevent such crises in the future and to enhance international cooperation.

The key theme of the report is the urgent need to move toward

> ❝ Humanity's inability to fit its activities into that pattern is changing planetary systems, fundamentally. Many such changes are accompanied by life-threatening hazards. This new reality, from which there is no escape, must be recognized—and managed. ❞
>
> World Commission on Environment and Development,* *Our Common Future*

sustainable development.* This theme arises from the understanding that growing human needs did not fit within the existing ecological system of the planet and that a major review of international policies of natural resource management was essential.

Exploring the Ideas

Our Common Future aims to convince the reader that sustainable development is of the utmost importance. The report demanded serious attention from politicians and the general public, arguing that immediate action should be taken to review unsustainable practices in managing natural resources.[2] It drew attention to major issues in contemporary human society, such as the importance of raising public awareness about environmental concerns. These concerns are organized into a set of concrete problems, referred to as challenges, which result from unsustainable trends. Later in the report, possible solutions are introduced and discussed.

Following a short introduction written by the commission chairperson Gro Brundtland that provides information about the commissioning of the report and the commissioners who worked on it, the main body of the report introduces the key argument immediately: "Hope for the future is conditional on decisive political action now to begin managing environmental resources to ensure both sustainable human progress and human survival."[3]

The text's opening overview introduces the global concerns and challenges of human development,* and then suggests possible solutions in the form of international cooperation and institutional reform. The global concerns presented in the main part of the report include the question of human survival in the future, the economic development triggered by poverty, crises, and economic decline, and the overexploitation of natural resources.[4] Global challenges, referred to as "common challenges," are named and discussed, such as the rapid growth of the human population, food insecurity,* climate change, the extinction of many species of flora and fauna, energy insecurity, the hazards of industrialization (the large-scale increase in industrial production), and urban development issues.[5]

The report's concluding discussion of possible solutions, called "common endeavors," suggests areas where international cooperation is needed, such as in the management of common global resources, like the oceans, outer space, and Antarctica, and the need for disarmament, clean energy, and institutional and legal actions to secure global peace and cooperation.[6]

Language and Expression

Although *Our Common Future* is a report commissioned by the United Nations* General Assembly,* its structure is quite different to the traditional format of UN reports that usually give brief background information and concentrate on proposed actions. It is also unusual for a UN report to be published as a book, as *Our Common Future* was.

The authors designed the report to be accessible to the widest possible audience, and aimed "to translate [our] words into a language that can reach the minds and hearts of people young and old."[7] Since the report was targeted at everyone, its argument had to be developed and presented in a simple but powerful way. As a result, it appears to be more comprehensive than other reports on the subject. It elaborates its main topic and attempts to convince readers that sustainable development has immense significance.

Our Common Future demands serious attention from a global audience and calls for immediate action at every level of human activity.[8] It expresses major concerns about contemporary human society, such as climate change, food insecurity, pollution, and a fast-growing population. These concerns are presented as a set of concrete problems, referred to as "challenges," that result from unsustainable general trends. Possible solutions are also introduced and discussed.

NOTES

1   World Commission on Environment and Development (WCED), *Our Common Future* (Oxford: Oxford University Press, 1989), xi.

2   WCED, *Our Common Future*, viii.

3   WCED, *Our Common Future*, 1.

4   WCED, *Our Common Future*, 27–36.

5   WCED, *Our Common Future*, 95–257.

6   WCED, *Our Common Future*, 261–343.

7   WCED, *Our Common Future*, xiv.

8   WCED, *Our Common Future*, x.

MODULE 6
SECONDARY IDEAS

KEY POINTS

- In its discussion of current challenges and possible solutions, *Our Common Future* used a "causal analysis"* to distinguish between the causes and consequences of the crises.

- This causal analysis was later developed into a new approach to policy* analysis, that of monitoring and evaluation*—a process in which an activity is constantly checked to see if it is effective, and information about it is continuously updated.

- Based on monitoring and evaluation, *Our Common Future* praises nongovernmental organizations (NGOs*) as a principle force in policy change. (NGOs are nonprofit organizations independent of the government of the nation in which they operate.)

Other Ideas

Applying causal analysis to policymaking by distinguishing between the causes and consequences of significant crises, as Gro Brundtland's *Our Common Future* did, was groundbreaking. Previously, it was mainly philosophers, not policymakers, who explained situations in terms of causes and effects. Calling for a deeper analysis, *Our Common Future* succeeded in influencing the approaches employed by other policymakers and NGOs in their management of environmental resources. The report inspired a great deal of research founded on monitoring and evaluation, according to which the researcher continuously assesses and tracks results. Since then, many economists, public managers, and biologists (those engaged in the study of living

> 66 We must understand better the symptoms of stress that confront us, we must identify the causes, and we must design new approaches to managing environmental resources and to sustaining human development. 99
>
> World Commission on Environment and Development,* *Our Common Future*

things) have developed methods to estimate the importance and impact of environmental and socioeconomic policies on ecosystems.[*1]

Brundtland and her commissioners believed in humankind's capacity to change behavioral patterns that negatively impact the environment, and to cooperate at all levels.[2] This belief, clearly conveyed in the report, can be partially explained by the moment in which the report was written and the aspirations of the generations that had lived through the Cold War* years of 1947 to 1991. It was a time of relative optimism and many people believed at that point that governments could work together to reshape the future.

Although the report also stimulated a change in the attitudes and political thinking of the era, the signs of change were often interpreted overly optimistically. As a result, complex issues were simplified into a common solution: cooperation between nations.

Exploring the Ideas

Although *Our Common Future* identified a link between the environment and human development,* this link required further study to understand existing patterns, to find tools to measure them, and to elaborate necessary policy interventions (changes in policy intended to produce a certain effect). Imbalanced approaches risked causing new crises rather than solving existing ones.

The report had the potential to bring about comprehensive systematic changes, but did not provide the methodological tools—

the specific, technical, methods—needed to help tackle actual crises. The double crisis of the natural environment and human development is complex, and it can be difficult to distinguish between causes and symptoms.[3] The report approached these complex issues in a very simple manner, only briefly discussing challenges before offering possible solutions.

The report's causal analysis was helpful in linking causes and their often-unintended effects in a new way. For example, the increasing use of raw materials and chemicals by industry has led to pollution; with a more comprehensive understanding of the causes of pollution, policymakers are better armed in their attempts to decrease its negative effects on human health and the natural environment.[4] Causal analysis also had an educational purpose, in that it helped to persuade the report's readers to adopt more sustainable practices as individuals.

Overlooked

While *Our Common Future* and its main ideas are well studied, its insights into the role of NGOs in promoting sustainable development* is still neglected in the current literature, with a few exceptions.[5] The relationship between NGOs and sustainable development is not so obvious. In fact, radical environmentalists such as the Indian activist Vandana Shiva* have even complained that sustainable development has been harmful for environmental movements, especially in poorer nations, because it risks killing the momentum of more radical struggles to protect the environment.[6] She was aware of the possibility that the introduction of "weak sustainability"* by Western international organizations would overshadow the "strong sustainability"* that environmental NGOs lobbied for. ("Weak sustainability" is a form of sustainability* that sees natural capital*—resources—as potentially replaceable by human capital*—knowledge, technology, and so on; "strong sustainability" is sustainability prioritizing the environment above all.)

Many NGOs contributed to the research for *Our Common Future*, submitting facts they had collected and making suggestions for policy changes. The authors knew it would not be easy for governments to accept policy changes, as the practicalities of sustainable development would require the fundamental reform of the management of resources, technology, institutions, and policy. Change like this is difficult to implement and demands institutional and financial resources, requiring politicians to make "painful choices."[7]

From the beginning, the authors of *Our Common Future* expected to encounter resistance to radical change. They expected that governments would have serious difficulties in addressing environmental crises, because their interests tended to be "too narrow, too concerned with quantities of production or growth."[8] In that sense, sustainable development is a "nongovernmental" topic—that is, governments will not necessarily be the best at making societies more sustainable, and need external pressure and encouragement. In Brundtland's vision, NGOs were pioneers in promoting public awareness and pushing governments to act.[9] However, the report's commissioners have never discussed the weaknesses of NGOs: that they tend to be focused on specific interests and have limited capacity to fund projects without government support.

NOTES

1   William R. Shadish et al., *Foundations of Program Evaluations: Theories of Practice* (London: Sage, 1991), 20–1.

2   World Commission on Environment and Development (WCED), *Our Common Future* (Oxford: Oxford University Press, 1989), ix.

3   David Wasdell, *Studies in Global Dynamics No. 7—Brundtland and Beyond: Towards a Global Process* (London: Urchin, 1987), 7–8.

4   WCED, *Our Common Future*, 2–3.

5   Ben Pile, "Wishing Greenpeace an Unhappy Birthday," *Spiked*, September 12, 2011, accessed February 2, 2016, http://www.spiked-online.com/newsite/article/11068#.VrCGxPkS-Uk.

6   Vandana Shiva, "The Greening of Global Reach," in *Global Ecology: A New Arena of Political Conflict*, ed. Wolfgang Sachs (London: Zed Books, 1993).

7   WCED, *Our Common Future*, 9.

8   WCED, *Our Common Future*, 328.

9   WCED, *Our Common Future*, 328.

MODULE 7
ACHIEVEMENT

KEY POINTS

- The importance of *Our Common Future* was recognized all over the world; its recommendations still dominate the political agenda of the United Nations.*

- The report expounded universal values, gaining the sympathy of a wide readership.

- The report has not, however, been directly responsible for any major changes in terms of policy* or in the behavior of individuals.

Assessing the Argument

Gro Brundtland's *Our Common Future*, as its title suggests, was designed to have a general appeal across nations and generations. It aimed to change international and national policies, and individual behavior that damages the environment, as part of "a global agenda for change."[1] In addition to the global character of the problems discussed in the report, the fact that the 21 commissioners came from different nations and were "unanimous in our conviction that the security, well-being, and very survival of the planet depend on such changes"[2] was seen as proof that the report was relevant throughout the world.

Our Common Future set out to address the major challenges faced by the global community in the 1980s. At the time of publication in 1987, the authors hoped that these challenges could be resolved by the year 2000.[3] However, the actions of policymakers and the general public in response to the report were not sufficient to achieve the target in time. This can be demonstrated by the fact that all the problems mentioned in the report, such as pollution, population

> 66 I remember I gave a kind of colder warning about the results so far, saying that, yes, there is progress in many fields, there is some progress in some fields, but also in several areas no progress at all. 99
>
> Gro Brundtland, *Interview with Peter Ocskay,* Baltic University, Uppsala, Sweden, 1997

growth, economic crises, poverty, and climate change,* are still facing us today and have even worsened over time. For this reason, the problems discussed in the report are very familiar to today's readers.

Our Common Future noted the global nature of the issues under discussion, and how they applied to both developing and developed countries—that is, both wealthy and impoverished countries.[4] However, the specifically limited capabilities of developing nations in areas such as knowledge and economic resources were also discussed and an argument was made in favor of increasing the aid given to impoverished countries by developed countries.[5] Some scholars and activists, among them the Swiss economist Matthias Finger,* criticized the distinction made in the report between wealthy developed countries and impoverished developing countries. They argued that making distinctions like this weakened and divided the environmental ("Green") movement.[6]

Achievement in Context

Just three years after the publication of *Our Common Future*, the Canadian minister of the environment Lucien Bouchard* stated that the concept of sustainable developmen* as introduced by the report "has changed forever the way we think about the environment."[7]

In the decades since publication, the concept of sustainable development has been widely accepted, and has become an important framework for policymakers. Not only did the idea of sustainable development produce the new discipline of environmental economics,*[8] it has also shaped the policymaking processes at all possible levels and become a common political term.

By 2009, the sustainable development agenda was incorporated into national development* strategies in 106 countries. The business sector had accepted a new agenda, corporate and social responsibility* (CSR), although some businesses who took this up were criticized for "greenwashing"—a marketing spin by businesses and organizations wanting to promote themselves as environmentally friendly.[9] Criticism of environmentally damaging business practices has led to some* forming joint initiatives to address sustainable development. National governments remain slow in reaching binding (compulsory) agreements about sharing responsibility for global matters, such as pollution, climate change, and food insecurity.*

The report's aim of promoting multilateral* collaboration, then, has partly been met. From this perspective, the report was a game changer for national and international policy, going far beyond the expectations of its authors.

Limitations

In the 1980s, the World Commission on Environment and Development* headed by Brundtland was hoping to achieve sustainable development by the year 2000.[10] However, despite all the actions suggested, the world has yet to solve the problems addressed in *Our Common Future* because "tensions, controversies and gridlocks between development and environment still exist."[11]

Despite worldwide recognition of the concept of sustainability,* little progress has been made in achieving it. The measures suggested by *Our Common Future* were voluntary.[12] Institutional follow-ups to the report, including the UN's Earth Summit* in 1992 in Rio de Janeiro and the Rio+20 Conference* in 2012, faced serious problems in "designing the move from theory to practice."[13] The main challenge is that "a huge constituency around the world cares deeply and talks about sustainable development, but has not taken serious on-the-ground action."[14] Similar feelings have been shared by policy analysts about the 1997 Kyoto Protocol* (an international treaty that aimed to slow climate change by reducing the emission of certain gases understood to cause it) and the Paris Conference* on climate change held in 2015.[15]

At a theoretical level, dissecting the concept of sustainable development into three parts (economic, social, and environmental development*) has led to another challenge: finding a balance between the environment and the economy. This challenge provokes fundamental debates on what should be prioritized: economic growth, as in the weak* sustainability model, or environmental conservation, as in the strong* sustainability model.[16]

NOTES

1 World Commission on Environment and Development (WCED), *Our Common Future* (Oxford: Oxford University Press, 1989), ix.

2 WCED, *Our Common Future*, 343.

3 WCED, *Our Common Future*, ix.

4 WCED, *Our Common Future*, 52.

5 WCED, *Our Common Future*, 60.

6 Matthias Finger, "Politics of the UNCED Process," in *Global Ecology: A New Arena of Political Conflict*, ed. Wolfgang Sachs (London: Zed Books, 1993), 36.

7 Linda Starke, *Signs of Hope: Working towards Our Common Future* (Oxford: Oxford University Press, 1990), 21.

8 Herman Daly, *Ecological Economics and Sustainable Development: Selected Essays* (Cheltenham: Edward Elgar, 2007), 251.

9 John Drexhage and Deborah Murphy, *Sustainable Development: From Brundtland to Rio 2012* (New York: United Nations, 2010), 15.

10 WCED, *Our Common Future*, ix.

11 Volker Hauff, "Brundtland Report: A 20 Years Update," keynote speech, European Sustainability, Berlin, June 3, 2007, accessed February 1, 2016, http://www.nachhaltigkeitsrat.de/uploads/media/ESB07_Keynote_speech_ Hauff_07-06-04_01.pdf.

12 Jennifer Elliott, *An Introduction to Sustainable Development: The Developing World* (London: Routledge, 2000), 8.

13 Elliott, *Introduction to Sustainable Development*, 8.

14 Drexhage and Murphy, *Sustainable Development*, 2.

15 Thomas Sterner, "The Paris Climate Change Conference Needs to Be More Ambitious," *Economist*, November 18, 2015, accessed February 2, 2016, http://www.economist.com/blogs/freeexchange/2015/11/cutting-carbon-emissions.

16 Drexhage and Murphy, *Sustainable Development*, 10.

MODULE 8
PLACE IN THE AUTHOR'S WORK

KEY POINTS

- Gro Brundtland made an important contribution to the discussion of contemporary global challenges, including environmental degradation, climate change,* and human health.

- *Our Common Future* provided a revolutionary approach to human development* in the 1980s and remains the best-known work on sustainability* despite recent criticisms; Gro Brundtland has continued to promote sustainable development.*

- *Our Common Future* is a milestone work in the careers of all the Brundtland commissioners,* including Brundtland herself.

Positioning

By the time Gro Brundtland was invited to chair the World Commission on Environment and Development,* she had already won a reputation as an experienced politician. Not only had she served two terms as the first female prime minister of Norway (1986–9 and 1990–6), she was also the only prime minister to have served as minister of the environment (1974–9).

This unique background allowed Brundtland to develop new ideas for the management of finite resources and the environment, and human development. She was already well known before the publication of *Our Common Future*, but the commission's report bolstered her status as a celebrity politician. She was subsequently invited to write short texts capturing major moments in the commission's history, including the foreword to a book about the

> ❝ In the end, I know I made the right choice of taking the challenge the UN gave me. It has had a great impact on this country [Norway] and globally. ❞
>
> Gro Brundtland, *Interview with the Uongozi Institute*

writing of the report, and she has given numerous public speeches. However, since then she has not produced a work equal to the significance of *Our Common Future*.[1]

In 1997–8, Brundtland wrote two autobiographical works in Norwegian, the titles of which translate as *My Life*[2] and *Dramatic Years 1986–1996*.[3] In 2002 she wrote an autobiography in English entitled *Madam Prime Minister: A Life in Power and Politics*.[4] The book describes major moments of her life and her encounters with other prominent leaders, such as the French president Francois Mitterand,* the French statesman Jacques Delors,* the first president of the Russian Federation Boris Yeltsin,* the US politician Hillary Clinton,* and the South African freedom fighter and president Nelson Mandela.* Her husband, Arne Olav Brundtland, also a Norwegian politician, wrote two memoirs capturing their lives during Gro Brundtland's public service. Their titles translate into English as *Married to Gro* and *Still Married to Gro*.[5] During Brundtland's tenure as director-general of the United Nations* health body, the World Health Organization* (WHO), she wrote about the influence of environmental factors and climate change on human health. Recently she has publicly discussed the role of democracy and human rights in achieving sustainable development.

Integration

Brundtland has remained in favor of *Our Common Future*'s main argument: that there is an important link between the management of environmental resources and economic growth. While she believes that the Northern hemisphere's development issues have almost been fixed, she has raised concerns about the mass poverty of developing countries, which has only seemed to worsen since the publication of the report. This point is also often raised by prominent critics of the report, such as the Indian environmental activist Vandana Shiva*[6] and the Swiss economist Matthias Finger.*[7] *Our Common Future* discussed the universal responsibility of humankind to nature and the necessity of cooperation between the Global North* and the Global South*; in her later writings Brundtland went even further in arguing that Northern consumers are partly responsible for the impoverishment and environmental degradation of the Global South, which could cause future conflicts.[8]

Brundtland used a similar approach to that of *Our Common Future* in her later works during her service in WHO. In 1998, she published "Macroeconomics and Health," with the economist Jeffrey D. Sachs,* which argued that health and longevity were both fundamental goals of development and the means to achieving development goals, and referenced the concept of sustainable development.[9] The WHO report also linked health to poverty reduction and long-term economic growth.

It can be argued that Brundtland has maintained the views she expressed in the report regarding economic growth as a means of achieving international development goals. The negative environmental impact of human development as a consequence of human economic activity was not discussed in the report—and nor has Brundtland mentioned it since.

Significance

Our Common Future is Brundtland's best-known work by far. When she began the process of chairing the commission that produced the report, she had already earned a good reputation in UN circles and this enhanced the commission's political credibility. The report made Brundtland internationally famous, as she was responsible for formulating arguably one of the most important concepts of modern political economy—sustainable development.

Brundtland, today a member of the Elders* (a think tank* made up of retired senior UN officials and former political leaders), has remained loyal to the ideas and ideals presented in *Our Common Future*. She has echoed them in her later, lesser-known works and has carried on promoting the concept of sustainable development in the years since the publication of the report, notably during her tenure as the director-general of WHO.

While the new ideas presented in *Our Common Future* were well received by most governments and by the general public in the 1980s,[10] more recent interpretations of the report and its core concept of sustainable development have been more critical. Environmental activists and political economists have referred to the report as an example of neoliberal* thinking, arguing that it relies on the false possibility of infinite economic growth based on efficient use of resources through technological advancement and international cooperation.[11] ("Neoliberalism" is an approach to economics that calls for the market to operate without hindrance from regulation or government intervention, and without regard to any social consequences.) Although Brundtland acknowledged and agreed with this critique in a public speech in Vienna in 2013, she has not published anything on the matter.

NOTES

1   See Linda Starke, *Signs of Hope: Working towards Our Common Future* (Oxford: Oxford University Press, 1990).

2   Gro Harlem Brundtland, *Mitt Liv: 1939–1986* (Oslo: Gylenhal, 1997).

3   Gro Harlem Brundtland, *Dramatiske ar, 1986–1996* (Oslo: Gyldedal, 1998).

4   Gro Harlem Brundtland, *Madam Prime Minister: A Life in Power and Politics* (New York: Farrar, Straus and Giroux, 2005), 19.

5   See Arne Olav Brundtland, *Gift med Gro* (Oslo: Oslo University, 1996) and Arne Olav Brundtland, *Fortsatt Gift med Gro* (Oslo: Oslo University, 2003).

6   Vandana Shiva, "The Greening of Global Reach," in *Global Ecology: A New Arena of Political Conflict*, ed. Wolfgang Sachs (London: Zed Books, 1993), 149–56.

7   Matthias Finger, "Politics of the UNCED Process," in *Global Ecology: A New Arena of Political Conflict*, ed. Wolfgang Sachs (London: Zed Books, 1993), 36.

8   Gro Harlem Brundtland, "Global Change and Our Common Future: The Benjamin Franklin Lecture," in *Global Change and Our Common Future*, ed. R. S. DeFries and T. F. Malone (Washington, DC: National Academy Press, 1989), 15.

9   Gro Harlem Brundtland and Jeffrey Sachs, *Macroeconomics and Health: Investing in Health for Economic Development* (Geneva: World Health Organization, 2001), 3.

10   Linda Starke, *Signs of Hope*, 2.

11   Gearoid O Tuathail et al., eds, *The Geopolitics Reader* (London: Routledge, 1998).

SECTION 3
IMPACT

MODULE 9
THE FIRST RESPONSES

KEY POINTS

- While *Our Common Future* and its main concept, sustainable development,* received a generally positive response, the main critiques came from environmental activists and economists who criticized the report for adopting an approach known as "weak* sustainability."

- None of the report's authors have entered into a public debate with their critics.

- Gro Brundtland's later works partly respond to criticisms by introducing a fourth element, culture, into the three-part model of sustainability given in the report.

Criticism

Although the recommendations for raising environmental awareness made in Gro Brundtland's *Our Common Future* were generally welcomed by scientists and policymakers,* some aspects of the new concept of sustainable development were criticized. One of the first critiques came from environmental activists and economists, who argued that the concept of sustainable development as described in the report was very broad and did not specify what exactly had to be sustained. In the report, sustainability was used both to refer to the responsible management of natural resources and as an economic term. For example, the authors used the concept of sustainability in relation to foreign debt, which was described as "unsustainable" when it results in the flight of local capital from developing countries (that is, when money from a less-wealthy nation is invested elsewhere).[1] Generally, in the report, sustainable development was equated with

> **❝** In my final response I just exclaimed: 'Yes, all is indeed linked to everything else.' At the time, 20 years ago, this final statement from me as prime minister was immediately criticized by the opposition, even ridiculed by some, for being evasive and unclear. Interestingly, today it is often quoted with great respect and even admiration. **❞**
>
> Gro Brundtland, *Climate Change and Our Common Future*, speech at the Global Climate and Energy Project (GCEP) Symposium

sustainable growth, promising economic prosperity under a new order of responsible management of natural resources.[2]

The Indian social scientist Shiv Visvanathan* has questioned how realistic it is to combine the concepts of environmental sustainability and development.* He argued that the type of development imposed by international organizations such as the United Nations* was "a genocidal act of control" forced upon developing countries and did not adequately address real ecological issues.[3]

The way that *Our Common Future* linked the state of the environment and economic development was innovative at the time. But ecological economists* termed this approach "weak sustainability," because it saw innovation and knowledge as a possible substitute for natural capital* (natural resources useful to human life and economics).[4] For them, the approach gave "a sense of comfort to the effect that the environment can be dispensed with." Instead they demanded "strong sustainability"—an approach that gave more emphasis to the protection of natural resources.[5]

Responses

Brundtland has maintained the position she expressed in the report about the need for sustainable development. Over the years she has deepened her understanding of the concept of sustainability by adding issues of democracy and human rights.[6] The report has been criticized as an example of how greenwashing* (superficial support for ecological issues without effecting real change, often in the name of material gain) can damage the global environmental movement, as argued by the Indian environmental activist Vandana Shiva.[7] However, neither Brundtland, nor any of the other commissioners who helped to produce the report, have engaged in a public discussion with their critics on this matter.

Our Common Future discussed the universal responsibility of humankind to protect nature and the necessity of cooperation between the Global North* and the Global South.* In her later writings, Brundtland went even further in arguing that Northern consumers played a role in the impoverishment of the Global South, which could serve as a source of future conflicts.[8] This could be seen as a response to her critics.

It is not known how much Brundtland agrees with the ecological economists who critiqued the concept of sustainable development, disagreeing with the report's argument that natural capital can be replaced by the knowledge, technology, and innovation known as "human capital." No one, it seems, has thought to ask her this question directly. She is known to be a cautious politician and is careful about the public platforms and writing opportunities she chooses.[9]

Conflict and Consensus

The concept of sustainability presented in *Our Common Future* was criticized by economists and environmental activists for promoting economic growth over environmental protection. It was also criticized for using a neoliberal* model of development in which the market must be allowed to operate without hindrance, and according to which the Global North dictates to the Global South, ignoring or destroying local environmental movements.[10]

However, the authors of the report believed that their main goal was to raise awareness about environmental and social crises, and not to recommend an exact set of actions. Like Brundtland, many of the authors were politicians themselves, and their status kept them from suggesting radical reforms to the world order. In fact, they considered the idea of sustainability to be radical enough and were aware of the political risks involved in pushing for further reforms.

Our Common Future and Brundtland's speeches indicate that the concept of sustainability was centered on people and their needs and she did not see anything wrong with that. Humanity and nature were seen as an interlinked system: "People influence the trends that [decide the destiny of] the planet. The planet affects people."[11] Brundtland believed that focusing on human needs was beneficial for both people and the planet.

The main compromise that Brundtland has made in response to critiques of the report was introducing a fourth element into the three-pillar system of sustainability: culture.[12] She added the concept of cultural relativism—the idea that different cultures have different values and practices that cannot be compared directly to each other—to the concept of sustainability. This allowed her to explain why different countries or regions introduced sustainability at different speeds and in different ways. This new addition avoided the one-size-fits-all approach for which the report had previously been criticized.

NOTES

1   World Commission on Environment and Development (WCED), *Our Common Future* (Oxford: Oxford University Press, 1989), 73.

2   WCED, *Our Common Future*, 68.

3   Shiv Visvanathan, "Mrs Brundtland's Disenchanted Cosmos," *Alternatives* 16 (1991): 378–81.

4   David Pearce and Giles Atkinson, "The Concept of Sustainable Development: An Evaluation of its Usefulness Ten years after Brundtland," *Swiss Journal of Economics and Statistics* 134, no. 3 (1998): 4–5.

5   P. A. Victor et al., "How Strong is Weak Sustainability?" *Economie Appliquée*
    48, no. 2 (1995): 75–94.

6   Gro Harlem Brundtland, *Opening Speech to the 2015 Nobel Peace Prize
    Forum*, Oslo, March 10, 2015, accessed February 1, 2016, https://www.
    youtube.com/watch?v=LBRmSjsnVGs.

7   Vandana Shiva, "The Greening of Global Reach," in *Global Ecology: A New
    Arena of Political Conflict*, ed. Wolfgang Sachs (London: Zed Books, 1993),
    149–56.

8   Gro Harlem Brundtland, "Global Change and Our Common Future: The
    Benjamin Franklin Lecture," in *Global Change and Our Common Future*, ed.
    R. S. DeFries and T. F. Malone (Washington, DC: National Academy Press,
    1989), 15.

9   David Wilsford, *Political Leaders of Contemporary Western Europe: A
    Biographical Dictionary* (Westport, CT: Greenwood Press, 1995), 55.

10 Visvanathan, "Mrs Brundtland's Disenchanted Cosmos," 378–9.

11 Gro Harlem Brundtland, "Healthy People, Healthy Planet," the Annual
    Lecture in the Business and the Environment Program, London, March 15,
    2001, accessed February 1, 2016, http://www.cisl.cam.ac.uk/publications/
    archive-publications/brundtland-paper.

12 Gro Harlem Brundtland, "ARA Lecture," Vienna Technical University,
    November 18, 2013, accessed February 1, 2016, https://www.youtube.
    com/watch?v=X7Z2o8tZZoE.

MODULE 10
THE EVOLVING DEBATE

KEY POINTS

- Environmental activists and economists have criticized *Our Common Future* and the concept of sustainability* for focusing too much on economic growth.

- United Nations (UN)* officials and the Brundtland commissioners* have recognized that progress toward achieving sustainability has been slow.

- A new generation of activists and progressive economists from developing nations have offered alternative frameworks to those in the report.

Uses and Problems

Our Common Future was researched and written collaboratively under the leadership of Gro Brundtland. Its central concept of sustainable development* has been at the core of contemporary debates on human development* ever since its publication; it continues to be the main framework for analysis of many contemporary global matters. The report remains useful in reminding current leaders about the complex nature of the link between the natural environment and international development.

In 2012, the UN held a conference on sustainable development in the Brazilian city of Rio de Janeiro, Rio+20,* at which national governments were offered a non-binding agreement known as *The Future We Want*. While some governments voluntarily committed to achieving sustainable development, as set out in the agreement, the document's lack of compulsory commitments led to widespread disappointment with the outcomes of the conference; media and civil

> ❝ While sustainable development is often perceived as an environmental issue, it has been subject to competing agendas. ❞
>
> John Drexhage and Deborah Murphy, *Sustainable Development: From Brundtland to Rio 2012*

society* saw it as "a caricature of diplomacy."[1]

The gravity of the current problems in international development and the lack of progress in tackling them have also shifted perspectives regarding *Our Common Future*. The general attitude toward the report and its optimistic account of the world's capacity to successfully address the economic, environmental, and social crises has changed from initial praise and optimism to more critical and sometimes pessimistic opinions. Some, such as the Slovenian social theorist Drago Kos,* have expressed strong doubts about the world's ability to ever achieve sustainability.[2]

Schools of Thought

Initially, all contributors to *Our Common Future* agreed about the need to manage environmental resources sustainably. Later, the concept of sustainable development was debated by proponents of strong* sustainability, with its emphasis on environmental protection, and others who argued for weak* sustainability, which acknowledged the importance of economic development.*[3] Supporters of the weak sustainability approach include Brundtland commissioners, such as the German politician Volker Hauff* and the Canadian businessman and diplomat Maurice Strong,* who argued that lack of economic growth has caused environmental deterioration. This approach is naturally supported by governments and corporations that tend to prioritize economic growth above all.

In contrast, environmental or ecological economists such as

Herman Daly,* a supporter of strong sustainability, argue that the protection of natural resources should be given priority and that doing otherwise will lead to further economic deterioration. The school of ecological economics has embraced the concepts of strong sustainability and weak sustainability and integrated them into its discussions regarding the place of natural capital* in the management of resources.[4]

In Current Scholarship

The arguments made by environmental economists and environmental activists about the benefits of strong sustainability over weak sustainability are still being debated.

The most recent version of the strong sustainability argument is an approach called "degrowth," which means rejecting ideas of neoliberal* economic growth. The concept was invented by grassroots environmental movements, particularly in South America, and was first presented in an academic framework by the Columbian American anthropologist Arturo Escobar.* The concept reformulates the meaning of economic growth. Under the degrowth model, the economy can be allowed to grow only in the sectors that benefit people and their livelihoods, such as health and education.[5] This model promotes the strong sustainability approach, as it argues for the protection of nature. An example of how this model is realized in practice is the Universal Declaration of the Rights of Mother Earth,[6] presented in Cochabamba, Bolivia, in April 2010. This was an alternative to the Earth Charter that failed to be adopted at the Earth Summit* in 1992.[7] Bolivia has pioneered the way in passing this document, which put the concept of the earth's rights on the UN agenda.

Ecological economists* also continue to criticize the model of economic growth based on production and consumption, especially when it overtakes the rate at which natural resources can replenish themselves. For example, the Harvard economist Michael Porter*[8] envisions a more efficient use of resources based on technological innovation, a win–win situation that would make the overexploitation of natural resources unnecessary. This notion is in line with *Our Common Future*, which considered technology one of the solutions to global challenges. Based on Porter's work, a new generation of economists[9] has offered a comprehensive framework to build a sustainable economy. This framework includes new business models based on ethical values and less focus on economic profit, promotes community–based ownership structures such as cooperatives (businesses or services owned by the people who work at them), and aims to achieve better measures of socioeconomic progress (that is, measures of when things are getting better socially and economically).

NOTES

1   Jim Leape, "It's Happening, but Not in Rio," *New York Times*, June 24, 2012, accessed February 1, 2016, http://www.nytimes.com/2012/06/25/opinion/action-is-happening-but-not-in-rio.html?_r=0.

2   Drago Kos, "Sustainable Development: Implementing Utopia?" *SOCIOLOGIJA* 54, no. 1 (2012): 7–20.

3   Jerome Pelenc, "Weak Sustainability Versus Strong Sustainability," Brief for *GSDR* (Louvain: UCL, 2015), 1–4.

4   John M. Gowdy and Marsha Walton, "Sustainability Concepts in Ecological Economics," *Economics Interactions with Other Disciplines*, vol. 2. Encyclopedia of Life Support Systems (Paris: UNESCO/Eolss, 2008), 111–20.

5   Arturo Escobar, "Alternatives to development," interview with Rob Hopkins, Venice, September 28, 2012, accessed February 1, 2016, http://transitionculture.org/2012/09/28/alternatives-to-development-an-interview-with-arturo-escobar/.

6   Global Alliance for the Rights of Nature, "Universal Declaration of Rights of Mother Earth," Cochabamba, Bolivia, April 22, 2010, accessed December 14, 2015, http://therightsofnature.org/universal-declaration/.

7   Maurice Strong, "The 1992 Earth Summit: An Inside View," interview with Philip Shabecoff, Quebec, 1999, accessed February 1, 2016, http://www.mauricestrong.net/index.php/earth-summit-strong.

8   Michael E. Porter and Claas van der Linde, "Green and Competitive: Ending the Stalemate," *Harvard Business Review* 73, no. 5 (1995).

9   D. W. O'Neill et al., "Enough Is Enough: Ideas for a Sustainable Economy in a World of Finite Resources," *Report of the Steady State Economy Conference* (Leeds: CASSE and Economic Justice for All, 2010), 9.

MODULE 11
IMPACT AND INFLUENCE TODAY

KEY POINTS

- *Our Common Future* remains significant today in both academic thinking and public policy.*
- The concept of sustainability* presented in *Our Common Future* has been criticized for being difficult to achieve in practice.
- Gro Brundtland still works in international politics as a senior policy adviser and remains committed to the concept of sustainable development.*

Position

Gro Brundtland's *Our Common Future* is the definitive source for the concept of sustainable development, and remains at the heart of current debates regarding the future path of development.*[1]

One major obstacle in the way of achieving sustainable development is how fragmentary the required actions are, involving not just changes in the behavior of national governments and how they work together, but in the behavior of individuals. Another is the ambiguity of the concept of sustainable development, and the complexity of the issues it covers.[2] Dealing with the future of human development, the topic has very high stakes; consequently, the debate about sustainable development has gone well beyond the narrow confines of those with an interest in the field of ecological economics. It is an issue that concerns us all.

The United Nations (UN),* the institution that originally commissioned the *Brundtland Report*, has never questioned the viability of sustainable development. Instead, it has tried to incorporate

> ❝ Sustainable development is a visionary development paradigm, and over the past 20 years governments, businesses, and civil society have accepted sustainable development as a guiding principle … ❞
>
> John Drexhage and Deborah Murphy, *Sustainable Development: From Brundtland to Rio 2012*

critiques and adapt the concept to new challenges. UN agencies took on the role of putting sustainable development into practice. The first institutional follow-up to the report was at the Earth Summit* in 1992 when UN officials, including Maurice Strong,* the secretary-general of the conference and a Brundtland commissioner,* tried to formulate the necessary policy changes in a concrete set of actions known as Agenda 21.* Another attempt at international agreement on sustainable development took place 20 years later in Rio de Janeiro at the Rio+20 Conference.* This resulted in an agreement that the Millennium Development Goals,* a list of aims for human development around the world, would include sustainability targets and be referred to as the Sustainable Development Goals (SDGs).*

Interaction

There have been several UN conferences to promote the implementation of sustainable development as recommended in *Our Common Future*. Of these, the Earth Summit of 1992 and the Rio+20 Conference of 2012 were particularly important. The summits were criticized for putting too much emphasis on environmental issues, and paying too little attention to development aid from wealthy to impoverished countries and to cooperation between countries.[3]

Gro Brundtland and Maurice Strong, both high-profile members of the Brundtland Commission, have remained engaged in the debate on sustainability for more than 20 years. Despite heavy criticism of the slow progress of sustainability, Brundtland and Strong stayed loyal to the concept, believing that the barriers to sustainability did not reflect fundamental flaws in the concept but were attributable to other factors.

Brundtland recognized that the challenge of sustainability was bigger than the system that was supposed to implement it. According to her recent speeches, international institutions could make better progress towards achieving it. Strong has also referred to institutional weaknesses of the UN that made it difficult for the organization to provide strong global leadership on sustainability. However, he was highly appreciative of the attempts made by Boutros Boutros-Ghali,* UN secretary-general between 1992 and 1996, to reform the UN, "bring[ing] together several of the elements within the secretariat into a new economic and social affairs department" called the Economic and Social Council (ECOSOC),* a body responsible for economic and social activities.[4] ECOSOC was given a central role in following up the decisions made at the Rio Conference.

Strong served as a senior adviser at the Rio+20 Conference, and used this opportunity to suggest institutional innovations to the UN system; among these were restructuring the United Nations Environment Program (UNEP),* the UN body responsible for coordinating the organization's environmental activities and policies, and offering assistance to developing nations seeking to implement practices and policies designed to protect the environment.[5] Despite criticisms, the Rio+20 Conference succeeded in reformulating the Millennium Development Goals as Sustainable Development Goals (SDGs), ensuring that sustainability targets were included in official international development aims. This was an important achievement.

The Continuing Debate

The debate on how to make sustainability work in practice continues. While some, among them the Slovenian social scientist Drago Kos,*[6] argue that the concept is utopian (that is, it belongs in an impossibly perfect world), most scientists and politicians assume that it is possible to make sustainable development a reality. Many others, including the South African religious leader and activist Desmond Tutu,* insist that extra effort has to be made if humanity is to survive in the near future.

UN officials and Brundtland commissioners are aware of the slow progress on implementing sustainability. In 2010, the UN High Level Panel on Global Sustainability, a committee of experts convened to discuss the subject, commissioned scientists from the International Institute of Sustainable Development* (an independent, nonprofit organization founded to promote human development by sustainable means) to study the causes of this slow progress and possible solutions. Certain development specialists argue for deep institutional changes in global economics, particularly in the area of global business, in order to make sustainable development happen.[7]

Brundtland, herself a member of the UN High Level Panel on Global Sustainability, has also addressed this question. In a public lecture at Vienna University of Technology she gave her answer to the following question: "Why are we not succeeding in changing our ways and building a sustainable future 20 years after Rio?"[8] She thinks that there are political, social, and technological reasons for the slow progress, and the inefficiency of global systems of government is, for her, the main reason that sustainable development has not been sufficiently widely adopted.[9]

NOTES

1   Volker Hauff, "Brundtland Report: A 20 Years Update," keynote speech, European Sustainability, Berlin, June 3, 2007, accessed February 1, 2016, http://www.nachhaltigkeitsrat.de/uploads/media/ESB07_Keynote_speech_Hauff_07–06–04_01.pdf.

2   John Drexhage and Deborah Murphy, *Sustainable Development: From Brundtland to Rio 2012* (New York: United Nations, 2010),16.

3   Maurice Strong, *Where on Earth Are We Going?* (Toronto: Vintage Canada, 2001), 78.

4   Maurice Strong, "The 1992 Earth Summit: An Inside View," interview with Philip Shabecoff, Quebec, 1999, accessed February 1, 2016, http://www.mauricestrong.net/index.php/earth-summit-strong.

5   Maurice Strong, "Statement by Maurice F. Strong delivered to the Special United Nations General Assembly Event on Rio+20," New York, October 25, 2011, accessed February 1, 2016, http://www.unep.org/environmentalgovernance/PerspectivesonRIO20/MauriceFStrong/tabid/55711/Default.aspx.

6   Drago Kos, "Sustainable Development: Implementing Utopia?" *SOCIOLOGIJA* 54, no. 1 (2012): 20.

7   Drexhage and Murphy, *Sustainable Development*, 19–20.

8   Gro Harlem Brundtland, "ARA Lecture," Vienna Technical University, November 18, 2013, accessed February 1, 2016, https://www.youtube.com/watch?v=X7Z2o8tZZoE.

9   Brundtland, "ARA Lecture."

MODULE 12
WHERE NEXT?

KEY POINTS

- It is very likely that the definition of sustainable development*
  established in *Our Common Future* will remain in use for the
  foreseeable future.

- It is also likely that the understanding of sustainability*
  founded on the three pillars of environment, human
  development,* and economic growth will be expanded into a
  multidimensional model.

- *Our Common Future* is a groundbreaking, highly praised
  text that has provided the most widely used definition of
  "sustainable development."

Potential

Humanity is likely to carry on facing the challenges identified by *Our
Common Future*, such as climate change,* overpopulation, and
decreasing natural resources. Sustainability will remain at the forefront
of the international political agenda, and it is likely that policymakers*
will continue to draw upon the classic definition of sustainability as
formulated by Brundtland.

The issues discussed in the report continue into the twenty-first
century, and many have worsened. It is expected that by 2050 the
world's population will reach 9.1 billion, and the main increase will
occur in developing countries.[1] In order to feed this population and
avoid food insecurity,* food production will have to increase by 70
percent.[2] Balancing this increased demand in food with rapidly
decreasing natural resources, such as fertile land and water, will be
difficult and will require a serious change in resource management

> 66 The fact is a compelling reality, and should not easily be dismissed. Since the answers to fundamental and serious concerns are not at hand, there is no alternative but to keep on trying to find them. 99

World Commission on Environment and Development,* *Our Common Future*

and the political will to bring this about. The solutions offered in *Our Common Future*, such as technological development and international cooperation, will definitely have to be applied.

Failure to find the right solution could lead to a serious global crisis that could threaten the very survival of the human species. As the South African religious leader and activist Desmond Tutu* put it, "We have only one world, and if we destroy it, we are done."[3]

Future Directions

Our Common Future proposed a revolutionary approach to the management of natural resources by connecting the environment with human development and adding the third element of economic growth. These are the three pillars that make up the concept of sustainable development that has served well in explaining the complexities of human development, although the concrete meaning of each of these pillars is still disputed.

To address this issue, a number of political economists and social scientists, including philosophers, have started to offer multidimensional models of sustainable development. For example, the Colombian American anthropologist Arturo Escobar* has offered an alternative development concept that "take[s] into account place-based models of nature, culture, and politics."[4] Culture is the fourth element to be added to the original three-part model of sustainability.

Another dimension that can be added to the model is global

governance—the means of political cooperation between governments. It is clear, even from the example of *Our Common Future* itself, that innovative ideas about the future of mankind will remain just ideas unless there are concrete mechanisms to make them a reality. The report rightly predicted that national governments would be reluctant to shift government policy toward sustainability, and emphasized the importance of lobbying from civil society*—the general public—to make it happen. However, there are gaps in the current understanding of how the role of civil society can be optimized in this process, so this question will continue to inspire theories and debates.

Summary

Our Common Future is a report prepared by the United Nations* World Commission on the Environment and Development* in 1987. It is also known as the *Brundtland Report*, named after the chairperson of the Commission, the former Norwegian prime minister Gro Harlem Brundtland. The report is considered the definitive source text for the concept of sustainable development. Although some earlier works mentioned a similar concept, it was *Our Common Future* that popularized the idea of a link between human development and natural resources, arguing for responsible management of natural resources in order to maintain them for future generations. This has become accepted today as the definition of sustainable development. It is frequently used in both academic work and political contexts.

Our Common Future introduced the conceptual cornerstones of contemporary development policies: the three pillars of economic growth, social development, and natural and ecological conservation. The report received much positive attention immediately after publication and was hailed by Oxford University Press as "the most important document of the decade on the future of the world." Indeed, it is possible to expand this point by claiming the report to be one of the most influential texts of the twentieth century. As the challenges and the possible solutions discussed within it have remained relevant, the report continues to influence political thinking in the twenty-first century.

Most international organizations, including the UN, continue to include sustainable development in their agenda. The recent formulation and adoption of the UN's Sustainable Development Goals,* updating the Millennium Development Goals* adopted in New York in 2000, is another example of how sustainability remains relevant in the twenty-first century. There is no doubt that *Our Common Future* will remain a highly influential text for decades to come.

NOTES

1   Food and Agriculture Organization (FAO), "How to Feed the World in 2050" (Rome: Food and Agriculture Organization, 2009), 2.

2   FAO, "How to Feed the World," 2.

3   Desmond Tutu, "Is Sustainable Development a Luxury We Can't Afford?," interview with the Elders, Cape Town, May 12, 2012, accessed February 1, 2016, https://www.youtube.com/watch?v=ArHey8SVJQE.

4   Arturo Escobar, "Culture Sits in Places: Reflections on Globalism and Subaltern Strategies of Localization," *Political Geography* 20 (2001): 139–74.

GLOSSARY

GLOSSARY OF TERMS

Agenda 21: an action plan to implement sustainable development principles at both international and national levels, developed by the UN Conference on the Environment and Development in Rio de Janeiro in 1992. The 21 in its title refers to the twenty-first century. It was a voluntary agreement. While European countries strongly supported the plan, the United States opposed it (which explains why it was never fully implemented).

Bhopal disaster: a gas leak at a pesticide factory in the Indian city of Bhopal in December 1984, it is considered the worst-ever industrial disaster. At least 3,787 deaths were confirmed in the Madhya Pradesh region where the accident happened.

Brandt Report: a report drafted by the Independent Commission on International Development, chaired by the West German political leader Willy Brandt, dealing with global economic development.

Brundtland Commission: formally titled the World Commission on Environment and Development, a body convened by the United Nations with the aim of fostering cooperation between nations to pursue sustainable development.

Causal analysis: a method of studying the causes of studied effects.

Chernobyl disaster: an industrial accident in April 1986 at a nuclear power-plant in what is today Ukraine, then the Soviet Union. It was the worst nuclear power-plant accident the world has known, affecting thousands of people.

Civil society: the part of society consisting of everything except the state and business sectors.

Climate change: patterns of change in global weather conditions resulting from natural and man-made causes, linked to global warming.

Club of Rome: an international think tank founded in 1968 at the Academia dei Lincei in Rome. Its members include international bureaucrats, political leaders, and prominent businessmen. The Club actively participates in debates on international development.

Cold War (1947–91): a period of political tension between the United States and the Soviet Union, and their respective allies. While the two countries never engaged in direct military conflict, they engaged in covert and proxy wars and espionage against one another.

Corporate and Social Responsibility (CSR): a system of self-regulation by corporations, based on ethical principles. It promotes a "triple bottom line" of people, planet, and profit.

Cuban missile crisis: a confrontation between the United States and the Soviet Union in October 1962, provoked by the installation of nuclear weapons in Cuba. This was the moment when the world came closest to a nuclear war.

Development: a network of political and social processes aimed at increasing people's life choices; mostly used as a term by governments and other organizations that aid, manage, and regulate societies.

Earth Summit: a United Nations conference held in Rio de Janeiro, in June 1992. It promoted the Rio Declaration on the Environment and Development and Agenda 21.

Economic and Social Council (ECOSOC): one of the principal bodies of the United Nations system, responsible for economic and social activities.

Ecosystem: a biological system comprising all the organisms that exist in a specific physical environment.

Elders: an independent group of retired world leaders who work together for peace and human rights, brought together by the South African statesman Nelson Mandela in 2007.

Environmental economics: the study of economic matters as they are affected by issues of sustainability, and of the economic consequences of policies designed to protect the environment.

Food insecurity: a lack of food supply leading to malnutrition, hunger, and starvation. It is a condition often present in developing countries.

General Assembly: one of the main organizations of the United Nations and the only one in which all member nations have equal representation.

Global North: a global socioeconomic and political category comprising North America and Western Europe.

Global South: a global socioeconomic and political category comprising Africa, South America, Asia, and the Middle East.

Greenwashing: a form of marketing in which an organization pretends to be environmentally friendly while not making any real changes.

Human capital: knowledge, technology, potential innovation, and human skills valuable to human prosperity.

International Institute of Sustainable Development: an independent, nonprofit organization founded to promote human development by sustainable means. The organization reports on international negotiations, encourages innovation and communication, and attempts to engage citizens, businesses, and policymakers in issues surrounding sustainability.

Kyoto Protocol: an international treaty adopted in Kyoto, Japan, in 1997 that aimed to reduce carbon dioxide (CO_2) emissions and slow climate change.

Millennium Development Goals: eight international development goals introduced in 2000 by the UN Millennium Summit, originally scheduled to be achieved by 2015. These goals include commitments to eradicate poverty, provide primary education, protect women's rights, decrease child mortality rates, improve maternal health, combat diseases such as HIV/AIDS, maintain environmental sustainability, and foster international partnerships globally.

Monitoring and evaluation: a process in project management in which an activity is constantly checked to see if it is effective, and information about it is continuously updated.

Multilateralism: a concept in international relations developed by Miles Kahler, promoting international governance by multiple states.

Natural capital: the supply of natural resources, including geology, air, water, soil, and living organisms. It can be used for economic activity.

Neoliberalism: a school of economic thought that emerged in the nineteenth century and promotes policies of economic liberalization and free trade.

Nongovernmental organization (NGO): a nonprofit voluntary group that is organized at local, national, or international level.

Norwegian Labor Party: a social-democratic political party founded in 1887 and still active in the political arena of Norway promoting socialists values.

Palme Commission: the Commission on Disarmament and Security, chaired by Swedish politician Olaf Palme, that produced the report *Common Security* in 1982.

Paris Conference: the United Nations Climate Change Conference held in Paris in December 2015; it concluded with the proposal of a global agreement to reduce climate change (the Paris Agreement).

Policy: a course of action followed by a government, a political party, or a company.

Rio+20 Conference: the United Nations Conference on Sustainable Development (UNCSD) held in Rio de Janeiro in 2002 to follow-up on the activities proposed by the 1992 Earth Summit.

Stockholm Conference: also known as the United Nations Conference on the Human Environment, this meeting was held in Stockholm, on June 5–16, 1972. The conference was an important step towards formulating the sustainable development concept, as it elaborated principles of actions to balance environmental challenges and socioeconomic development, known as the Stockholm Declaration. The idea to establish the United Nations Environment Program (UNEP) was also put forward at the conference.

Soviet Union: the Union of Soviet Socialist Republics (USSR) was a socialist state that existed between 1922 and 1991, centered on Russia. It was a one-party state, governed by the Communist Party, with Moscow as its capital.

Special envoy: a person chosen by the UN secretary-general to act as a representative and adviser on a particular matter, such as human rights or climate change. The position is honorary and unpaid.

Strong sustainability: the idea supported by ecological environmentalists that human capital and natural capital may be complementary, but they are not interchangeable.

Sustainability: the main principle of responsible use and management of resources, presented in *Our Common Future*: a development model that "meets the needs of the present without compromising the ability of future generations to meet their own needs."

Sustainable development: the various policies and strategies designed to improve the lives of people while ensuring that future generations will be able to meet their needs.

Sustainable Development Goals (SDGs): a set of 17 goals to improve the condition of living globally. The idea of SDGs was discussed in 2012 at the Rio+20 Conference and accepted by the UN General Assembly in 2014.

Think tank: a group of experts providing advice and ideas on specific political or economic problems.

United Nations (UN): an intergovernmental organization created in 1945 by 51 founding member states to promote international cooperation and global peace.

United Nations Environment Program (UNEP): a UN agency specializing in environmental activities. It was created in 1972 on the initiative of Maurice Strong. Its headquarters are in Nairobi, Kenya.

Weak sustainability: an idea supported by environmental economics that contends that human capital and natural capital are interchangeable.

World Commission on Environment and Development: the original title of the Brundtland Commission, convened by the United Nations with the aim of fostering cooperation between nations to pursue sustainable development.

World Health Organization (WHO): one of the United Nations agencies specializing in global public health. Its headquarters are based in Geneva.

Yugoslavia: a European country that existed from 1918 until 1991. Its territory included what is today Bosnia and Herzegovina, Croatia, Macedonia, Montenegro, Slovenia, and Serbia.

PEOPLE MENTIONED IN THE TEXT

Lucien Bouchard (b. 1938) is a former minister of the environment in the Canadian government. A politician and diplomat, he was premier of the Canadian region of Quebec between 1996 and 2001.

Boutros Boutros-Ghali (1922–2016) was an Egyptian politician who served as secretary-general of the United Nations from 1992 to 1996.

Willy Brandt (1913–92) was chancellor of the Federal Republic of Germany from 1969 to 1974. In 1980 he chaired an Independent Commission that produced the *Brandt Report* on global economic development.

Lester R. Brown (b. 1934) is an American environmentalist and the author of several books discussing major challenges of the contemporary world and different scenarios of the human future. He is a founder of the think tank Worldwatch Institute and has written or coauthored more than 50 books, including *Man, Land and Food* (1963).

Hillary Clinton (b. 1947) is an American politician who served as secretary of state under President Barack Obama. She is a candidate for United States president in 2016.

Herman Daly (b. 1938) is an American professor of economics at the School of Public Policy of the University of Maryland who specializes in environmental economics.

Jacques Delors (b. 1925) is a French economist who served as the president of the European Commission between 1985 and 1995.

Nitin Desai (b. 1941) is an Indian economist and, as a member of the Brundtland Commission, an author of *Our Common Future*. He also served as under-secretary-general for economic and social affairs of the United Nations from 1992 to 2003.

Arturo Escobar (b. 1952) is a Columbian American anthropologist who studies social movements and international development. He is the Kenan Distinguished Professor of Anthropology at the University of North Carolina at Chapel Hill.

Matthias Finger (b. 1955) is the dean of the School of Continuing Education in the College of Management of Technology at the Swiss Federal Institute of Technology in Lausanne.

Volker Hauff (b. 1940) is a German politician; he is a member of the Social Democratic Party.

Wes Jackson (b. 1936) is one of the founders and leaders of sustainable agriculture movements. He has studied the relationship between human agricultural activity and ecosystems, and argued for more careful use of technology and nonrenewable resources.

Ban Ki-moon (b. 1944) is a former South Korean minister of foreign affairs and trade, and the current secretary-general of the United Nations.

Drago Kos (b. 1961) is a Slovenian lawyer, journalist, and former police officer. He was the first president of the OECD Working Group on Bribery.

Thomas Robert Malthus (1766–1834) was an English cleric and scholar who discussed the connection between famine and population growth, known as a Malthusian catastrophe. He published his most significant work, *An Essay on the Principle of Population*, in 1798.

Nelson Rolihlahla Mandela (1918–2013) was a South African human rights activist who served as the president of South Africa from 1994 to 1999.

Francois Maurice Adrien Marie Mitterand (1916–96) was a leader of the Socialist Party of France who served as the president of France from 1981 to 1995.

Saburo Okita (1914–93) was a Japanese economist and a minister of foreign affairs (1979–80) who chaired research on Japanese economic growth.

Sven Olaf Joachim Palme (1927–86) was a Swedish prime minister. He was chairman of the Commission on Disarmament and Security, also known as the Palme Commission, that produced the report *Common Security* in 1982.

Javier Perez de Cuellar (b. 1920) is a former prime minister of Peru and the fifth secretary-general of the United Nations. He commissioned Gro Brundtland to chair the World Commission on Environment and Development and write *Our Common Future* (the *Brundtland Report*).

Michael Porter (b. 1947) is an American economist who writes on corporate social responsibility and green business. He is the Bishop William Lawrence University Professor at the Institute for Strategy and Competitiveness at the Harvard Business School. He developed the "five forces analysis," a framework for analyzing the level of competition within an industry and business development.

Jeffrey Sachs (b. 1954) is an American economist known as an expert in development and poverty, and a special adviser to the secretary-general of the United Nations on sustainable development. He is the author of *The End of Poverty* (2005), *Common Wealth* (2008), and *The Price of Civilization* (2011).

Vandana Shiva (b. 1952) is an Indian environmentalist, feminist, and international activist. She leads campaigns against genetic engineering and works closely with grassroots organizations in developing countries.

Maurice Strong (1929–2015) was a Canadian businessman, the first director of UNEP, and one of the Brundtland commissioners who wrote *Our Common Future.*

Desmond Mpilo Tutu (b. 1931) is a retired bishop and a human rights activist who fought against racial discrimination in South Africa.

Shiv Visvanathan is an Indian public intellectual and social scientist.

Boris Yeltsin (1931–2007) was the first president of the Russian Federation.

WORKS CITED

WORKS CITED

Brown, Lester R. *Man, Land and Food: Looking Ahead at World Food Needs*. Washington, DC: US Department of Agriculture, 1963.

Brundtland, Arne Olav. *Gift med Gro*. Oslo: Oslo University, 1996.

———. *Fortsatt Gift med Gro*. Oslo: Oslo University, 2003.

Brundtland, Gro Harlem. "Global Change and Our Common Future: The Benjamin Franklin Lecture." In *Global Change and Our Common Future,* edited by R. S. DeFries and T. F. Malone. Washington, DC: National Academy Press, 1989.

———. *Interview with Peter Ocskay*. Baltic University, Uppsala, 1997. Accessed February 1, 2016. https://www.youtube.com/watch?v=ogrcy8AY95I.

———. *Mitt Liv: 1939–1986*. Oslo: Gylenhal, 1997.

———. *Dramatiske ar, 1986–1996*. Oslo: Gyldedal, 1998.

———. "Healthy People, Healthy Planet." The Annual Lecture in the Business and the Environment Program. London, March 15, 2001. Accessed February 1, 2016. http://www.cisl.cam.ac.uk/publications/archive-publications/brundtland-paper.

———. *Madam Prime Minister: A Life in Power and Politics*. New York: Farrar, Straus and Giroux, 2005.

———. *Interview with the Uongozi Institute*. Oslo, July 5, 2013. Accessed February 1, 2016. https://www.youtube.com/watch?v=8POtjDtH6io.

———. "ARA Lecture." Vienna Technical University. November 18, 2013. Accessed February 1, 2016. https://www.youtube.com/watch?v=X7Z2o8tZZoE.

———. *Opening Speech to the 2015 Nobel Peace Prize Forum*. Oslo, March 10, 2015. Accessed February 1, 2016. https://www.youtube.com/watch?v=LBRmSjsnVGs.

Brundtland, Gro Harlem, and Jeffrey Sachs. *Macroeconomics and Health: Investing in Health for Economic Development.* Geneva: World Health Organization, 2001.

Chatterjee, Pratap, and Matthias Finger. *The Earth Brokers: Power, Politics and World Development*. London: Routledge, 2013.

Daly, Herman. *Ecological Economics and Sustainable Development: Selected Essays*. Cheltenham: Edward Elgar, 2007.

Desai, Nitin. "Symposium: The Road from Johannesburg." Keynote Address. Georgetown: *Environmental Law Review*, 2003.

Dodds, Felix, and Michael Strauss with Maurice F. Strong. *Only One Earth: The Long Road via Rio to Sustainable Development*. London: Routledge, 2012.

Drexhage, John, and Deborah Murphy. *Sustainable Development: From Brundtland to Rio 2012*. New York: United Nations, 2010.

Elliott, Jennifer. *An Introduction to Sustainable Development: The Developing World*. London: Routledge, 2000.

Escobar, Arturo. "Culture Sits in Places: Reflections on Globalism and Subaltern Strategies of Localization." *Political Geography* 20 (2001): 139–74.

— — —. "Alternatives to Development." Interview with Rob Hopkins, Venice, September 28, 2012. Accessed February 1, 2016. http://transitionculture. org/2012/09/28/alternatives-to-development-an-interview-with-arturo-escobar/.

Food and Agriculture Organization. *How to Feed the World in 2050*. Rome: Food and Agriculture Organization, 2009.

Finger, Matthias. "Politics of the UNCED Process." In *Global Ecology: A New Arena of Political Conflict*, edited by Wolfgang Sachs. London: Zed Books, 1993.

Global Alliance for the Rights of Nature. "Universal Declaration of Rights of Mother Earth." Cochabamba, Bolivia, April 22, 2010. Accessed December 14, 2015. http://therightsofnature.org/universal-declaration/.

Gowdy, John M., and Marsha Walton. "Sustainability Concepts in Ecological Economics." *Economics Interactions with Other Disciplines, Volume 2*. Encyclopedia of Life Support Systems. Paris: UNESCO/Eolss, 2008.

Hauff, Volker. "Brundtland Report: A 20 Years Update." Keynote speech, European Sustainability, Berlin, June 3, 2007. Accessed February 2, 2016. http://www.nachhaltigkeitsrat.de/uploads/media/ESB07_Keynote_speech_ Hauff_07–06–04_01.pdf.

Jackson, Wes. *New Roots for Agriculture*. Lincoln: University of Nebraska Press, 1985.

Kos, Drago. "Sustainable Development: Implementing Utopia?" *SOCIOLOGIJA* 54, no. 1 (2012).

Leape, Jim. "It's Happening, but Not in Rio." *New York Times*, June 24, 2012. Accessed February 1, 2016. http://www.nytimes.com/2012/06/25/opinion/ action-is-happening-but-not-in-rio.html?_r=0.

Malthus, Thomas R. *An Essay on the Principle of Population*. London: J. Johnson, 1798.

Meadows, Donella H., Dennis L. Meadows, Jorgen Randers, and William W. Behrens III. *The Limits to Growth: A Report for the Club of Rome's Project on the Predicament of Mankind*. New York: Universe Books, 1974.

Norgard, Jorgen, John Peet, and Kristin Ragnarsdottir. "The History of the Limits to Growth." *Solutions* 2, no. 1 (2010): 59–63.

O'Neill, D. W., R. Dietz, and N. Jones. "Enough Is Enough: Ideas for a Sustainable Economy in a World of Finite Resources." *The Report of the Steady State Economy Conference.* Leeds: CASSE and Economic Justice for All, 2010.

O Tuathail, Gearoid, Simon Dalby and Paul Routledge (eds). *The Geopolitics Reader.* London: Routledge, 1998.

Parenivel, Mauree. "Sustainability and Sustainable Development." *Le Mauricien*, August 17, 2011. Accessed February 1, 2016. http://www. lemauricien.com/article/maurice-ile-durable-sustainability-and-sustainable-development.

Pearce, David, and Giles Atkinson. "The Concept of Sustainable Development: An Evaluation of its Usefulness Ten Years after Brundtland." *Swiss Journal of Economics and Statistics* 134, no. 3 (1998).

Pelenc, Jerome. "Weak Sustainability versus Strong Sustainability." Brief for *GSDR*. Louvain: UCL, 2015.

Pereira, Winin, and Jeremy Seabrook. *Asking the Earth: Farms, Forests and Survival in India.* Sterling, VA: Earthscan, 1990.

Pile, Ben. "Wishing Greenpeace an Unhappy Birthday." *Spiked*, September 12, 2011. Accessed February 2, 2016. http://www.spiked-online.com/newsite/article/11068#.VrCGxPkS-Uk.

Porter, Michael E., and Claas van der Linde. "Green and Competitive: Ending the Stalemate." *Harvard Business Review* 73, no. 5 (1995).

Shadish, William R., Thomas D. Cook, and Laura C. Leviton. *Foundations of Program Evaluations: Theories of Practice.* London: Sage, 1991.

Shiva, Vandana. "The Greening of Global Reach." In *Global Ecology: A New Arena of Political Conflict*, edited by Wolfgang Sachs. London: Zed Books, 1993.

Starke, Linda. *Signs of Hope: Working towards Our Common Future.* Oxford: Oxford University Press, 1990.

Sterner, Thomas. "The Paris Climate Change Conference Needs to Be More Ambitious." *Economist*, November 18, 2015. Accessed February 2, 2016. http://www.economist.com/blogs/freeexchange/2015/11/cutting-carbon-emissions.

Strong, Maurice F. "The 1992 Earth Summit: An Inside View." Interview with Philip Shabecoff, Quebec, 1999. Accessed February 1, 2016. http://www. mauricestrong.net/index.php/earth-summit-strong.

————. *Where on Earth Are We Going?* Toronto: Vintage Canada, 2001.

————. "Statement by Maurice F. Strong Delivered to the Special United Nations General Assembly Event on Rio+20." New York, October 25, 2011. Accessed February 1, 2016. http://www.unep.org/environmentalgovernance/PerspectivesonRIO20/MauriceFStrong/tabid/55711/Default.aspx.

Turner, Graham. *A Comparison of Limits to Growth with Thirty Years of Reality.* CSIRO Working Paper. Canberra: CSIRO, 2008.

Tutu, Desmond. "Is Sustainable Development a Luxury We Can't Afford?" Interview with the Elders, Cape Town, May 12, 2012. Accessed February 1, 2016. https://www.youtube.com/watch?v=ArHey8SVJQE.

United Nations. *Biography of Dr Gro Harlem Brundtland.* Geneva: UN, 2014.

Victor, P. A., J. E. Hanna, and A. Kubursi. "How Strong is Weak Sustainability?" *Economie Appliquée* 48, no. 2 (1995): 75–94.

Visvanathan, Shiv. "Mrs Brundtland's Disenchanted Cosmos." *Alternatives* 16 (1991): 378–81.

Wasdell, David. *Studies in Global Dynamics No.7—Brundtland and Beyond: Towards a Global Process.* London: Urchin: 1987.

Wilsford, David. *Political Leaders of Contemporary Western Europe: A Biographical Dictionary.* Westport, CT: Greenwood Press, 1995.

World Commission on Environment and Development. *Report of the World Commission on Environment and Development: Our Common Future.* August 4,1987. Accessed February 1, 2016. http://www.un-documents.net/wced-ocf.htm.

————. *Our Common Future.* Oxford: Oxford University Press, 1989.

World Health Organization. "Dr Gro Harlem Brundtland, Director-General." Geneva: WHO, 1998. Accessed March 9, 2016. http://www.who.int/dg/brundtland/bruntland/en/.

THE MACAT LIBRARY
BY DISCIPLINE

AFRICANA STUDIES

Chinua Achebe's *An Image of Africa: Racism in Conrad's Heart of Darkness*
W. E. B. Du Bois's *The Souls of Black Folk*
Zora Neale Huston's *Characteristics of Negro Expression*
Martin Luther King Jr's *Why We Can't Wait*
Toni Morrison's *Playing in the Dark: Whiteness in the American Literary Imagination*

ANTHROPOLOGY

Arjun Appadurai's *Modernity at Large: Cultural Dimensions of Globalisation*
Philippe Ariès's *Centuries of Childhood*
Franz Boas's *Race, Language and Culture*
Kim Chan & Renée Mauborgne's *Blue Ocean Strategy*
Jared Diamond's *Guns, Germs & Steel: the Fate of Human Societies*
Jared Diamond's *Collapse: How Societies Choose to Fail or Survive*
E. E. Evans-Pritchard's *Witchcraft, Oracles and Magic Among the Azande*
James Ferguson's *The Anti-Politics Machine*
Clifford Geertz's *The Interpretation of Cultures*
David Graeber's *Debt: the First 5000 Years*
Karen Ho's *Liquidated: An Ethnography of Wall Street*
Geert Hofstede's *Culture's Consequences: Comparing Values, Behaviors, Institutes and Organizations across Nations*
Claude Lévi-Strauss's *Structural Anthropology*
Jay Macleod's *Ain't No Makin' It: Aspirations and Attainment in a Low-Income Neighborhood*
Saba Mahmood's *The Politics of Piety: The Islamic Revival and the Feminist Subject*
Marcel Mauss's *The Gift*

BUSINESS

Jean Lave & Etienne Wenger's *Situated Learning*
Theodore Levitt's *Marketing Myopia*
Burton G. Malkiel's *A Random Walk Down Wall Street*
Douglas McGregor's *The Human Side of Enterprise*
Michael Porter's *Competitive Strategy: Creating and Sustaining Superior Performance*
John Kotter's *Leading Change*
C. K. Prahalad & Gary Hamel's *The Core Competence of the Corporation*

CRIMINOLOGY

Michelle Alexander's *The New Jim Crow: Mass Incarceration in the Age of Colorblindness*
Michael R. Gottfredson & Travis Hirschi's *A General Theory of Crime*
Richard Herrnstein & Charles A. Murray's *The Bell Curve: Intelligence and Class Structure in American Life*
Elizabeth Loftus's *Eyewitness Testimony*
Jay Macleod's *Ain't No Makin' It: Aspirations and Attainment in a Low-Income Neighborhood*
Philip Zimbardo's *The Lucifer Effect*

ECONOMICS

Janet Abu-Lughod's *Before European Hegemony*
Ha-Joon Chang's *Kicking Away the Ladder*
David Brion Davis's *The Problem of Slavery in the Age of Revolution*
Milton Friedman's *The Role of Monetary Policy*
Milton Friedman's *Capitalism and Freedom*
David Graeber's *Debt: the First 5000 Years*
Friedrich Hayek's *The Road to Serfdom*
Karen Ho's *Liquidated: An Ethnography of Wall Street*

John Maynard Keynes's *The General Theory of Employment, Interest and Money*
Charles P. Kindleberger's *Manias, Panics and Crashes*
Robert Lucas's *Why Doesn't Capital Flow from Rich to Poor Countries?*
Burton G. Malkiel's *A Random Walk Down Wall Street*
Thomas Robert Malthus's *An Essay on the Principle of Population*
Karl Marx's *Capital*
Thomas Piketty's *Capital in the Twenty-First Century*
Amartya Sen's *Development as Freedom*
Adam Smith's *The Wealth of Nations*
Nassim Nicholas Taleb's *The Black Swan: The Impact of the Highly Improbable*
Amos Tversky's & Daniel Kahneman's *Judgment under Uncertainty: Heuristics and Biases*
Mahbub Ul Haq's *Reflections on Human Development*
Max Weber's *The Protestant Ethic and the Spirit of Capitalism*

FEMINISM AND GENDER STUDIES

Judith Butler's *Gender Trouble*
Simone De Beauvoir's *The Second Sex*
Michel Foucault's *History of Sexuality*
Betty Friedan's *The Feminine Mystique*
Saba Mahmood's *The Politics of Piety: The Islamic Revival and the Feminist Subjec*t
Joan Wallach Scott's *Gender and the Politics of History*
Mary Wollstonecraft's *A Vindication of the Rights of Woman*
Virginia Woolf's *A Room of One's Own*

GEOGRAPHY

The Brundtland Report's *Our Common Future*
Rachel Carson's *Silent Spring*
Charles Darwin's *On the Origin of Species*
James Ferguson's *The Anti-Politics Machine*
Jane Jacobs's *The Death and Life of Great American Cities*
James Lovelock's *Gaia: A New Look at Life on Earth*
Amartya Sen's *Development as Freedom*
Mathis Wackernagel & William Rees's *Our Ecological Footprint*

HISTORY

Janet Abu-Lughod's *Before European Hegemony*
Benedict Anderson's *Imagined Communities*
Bernard Bailyn's *The Ideological Origins of the American Revolution*
Hanna Batatu's *The Old Social Classes And The Revolutionary Movements Of Iraq*
Christopher Browning's *Ordinary Men: Reserve Police Batallion 101 and the Final Solution in Poland*
Edmund Burke's *Reflections on the Revolution in France*
William Cronon's *Nature's Metropolis: Chicago And The Great West*
Alfred W. Crosby's *The Columbian Exchange*
Hamid Dabashi's *Iran: A People Interrupted*
David Brion Davis's *The Problem of Slavery in the Age of Revolution*
Nathalie Zemon Davis's *The Return of Martin Guerre*
Jared Diamond's *Guns, Germs & Steel: the Fate of Human Societies*
Frank Dikotter's *Mao's Great Famine*
John W Dower's *War Without Mercy: Race And Power In The Pacific War*
W. E. B. Du Bois's *The Souls of Black Folk*
Richard J. Evans's *In Defence of History*
Lucien Febvre's *The Problem of Unbelief in the 16th Century*
Sheila Fitzpatrick's *Everyday Stalinism*

Eric Foner's *Reconstruction: America's Unfinished Revolution, 1863-1877*
Michel Foucault's *Discipline and Punish*
Michel Foucault's *History of Sexuality*
Francis Fukuyama's *The End of History and the Last Man*
John Lewis Gaddis's *We Now Know: Rethinking Cold War History*
Ernest Gellner's *Nations and Nationalism*
Eugene Genovese's *Roll, Jordan, Roll: The World the Slaves Made*
Carlo Ginzburg's *The Night Battles*
Daniel Goldhagen's *Hitler's Willing Executioners*
Jack Goldstone's *Revolution and Rebellion in the Early Modern World*
Antonio Gramsci's *The Prison Notebooks*
Alexander Hamilton, John Jay & James Madison's *The Federalist Papers*
Christopher Hill's *The World Turned Upside Down*
Carole Hillenbrand's *The Crusades: Islamic Perspectives*
Thomas Hobbes's *Leviathan*
Eric Hobsbawm's *The Age Of Revolution*
John A. Hobson's *Imperialism: A Study*
Albert Hourani's *History of the Arab Peoples*
Samuel P. Huntington's *The Clash of Civilizations and the Remaking of World Order*
C. L. R. James's *The Black Jacobins*
Tony Judt's *Postwar: A History of Europe Since 1945*
Ernst Kantorowicz's *The King's Two Bodies: A Study in Medieval Political Theology*
Paul Kennedy's *The Rise and Fall of the Great Powers*
Ian Kershaw's *The "Hitler Myth": Image and Reality in the Third Reich*
John Maynard Keynes's *The General Theory of Employment, Interest and Money*
Charles P. Kindleberger's *Manias, Panics and Crashes*
Martin Luther King Jr's *Why We Can't Wait*
Henry Kissinger's *World Order: Reflections on the Character of Nations and the Course of History*
Thomas Kuhn's *The Structure of Scientific Revolutions*
Georges Lefebvre's *The Coming of the French Revolution*
John Locke's *Two Treatises of Government*
Niccolò Machiavelli's *The Prince*
Thomas Robert Malthus's *An Essay on the Principle of Population*
Mahmood Mamdani's *Citizen and Subject: Contemporary Africa And The Legacy Of Late Colonialism*
Karl Marx's *Capital*
Stanley Milgram's *Obedience to Authority*
John Stuart Mill's *On Liberty*
Thomas Paine's *Common Sense*
Thomas Paine's *Rights of Man*
Geoffrey Parker's *Global Crisis: War, Climate Change and Catastrophe in the Seventeenth Century*
Jonathan Riley-Smith's *The First Crusade and the Idea of Crusading*
Jean-Jacques Rousseau's *The Social Contract*
Joan Wallach Scott's *Gender and the Politics of History*
Theda Skocpol's *States and Social Revolutions*
Adam Smith's *The Wealth of Nations*
Timothy Snyder's *Bloodlands: Europe Between Hitler and Stalin*
Sun Tzu's *The Art of War*
Keith Thomas's *Religion and the Decline of Magic*
Thucydides's *The History of the Peloponnesian War*
Frederick Jackson Turner's *The Significance of the Frontier in American History*
Odd Arne Westad's *The Global Cold War: Third World Interventions And The Making Of Our Times*

LITERATURE

Chinua Achebe's *An Image of Africa: Racism in Conrad's Heart of Darkness*
Roland Barthes's *Mythologies*
Homi K. Bhabha's *The Location of Culture*
Judith Butler's *Gender Trouble*
Simone De Beauvoir's *The Second Sex*
Ferdinand De Saussure's *Course in General Linguistics*
T. S. Eliot's *The Sacred Wood: Essays on Poetry and Criticism*
Zora Neale Huston's *Characteristics of Negro Expression*
Toni Morrison's *Playing in the Dark: Whiteness in the American Literary Imagination*
Edward Said's *Orientalism*
Gayatri Chakravorty Spivak's *Can the Subaltern Speak?*
Mary Wollstonecraft's *A Vindication of the Rights of Women*
Virginia Woolf's *A Room of One's Own*

PHILOSOPHY

Elizabeth Anscombe's *Modern Moral Philosophy*
Hannah Arendt's *The Human Condition*
Aristotle's *Metaphysics*
Aristotle's *Nicomachean Ethics*
Edmund Gettier's *Is Justified True Belief Knowledge?*
Georg Wilhelm Friedrich Hegel's *Phenomenology of Spirit*
David Hume's *Dialogues Concerning Natural Religion*
David Hume's *The Enquiry for Human Understanding*
Immanuel Kant's *Religion within the Boundaries of Mere Reason*
Immanuel Kant's *Critique of Pure Reason*
Søren Kierkegaard's *The Sickness Unto Death*
Søren Kierkegaard's *Fear and Trembling*
C. S. Lewis's *The Abolition of Man*
Alasdair MacIntyre's *After Virtue*
Marcus Aurelius's *Meditations*
Friedrich Nietzsche's *On the Genealogy of Morality*
Friedrich Nietzsche's *Beyond Good and Evil*
Plato's *Republic*
Plato's *Symposium*
Jean-Jacques Rousseau's *The Social Contract*
Gilbert Ryle's *The Concept of Mind*
Baruch Spinoza's *Ethics*
Sun Tzu's *The Art of War*
Ludwig Wittgenstein's *Philosophical Investigations*

POLITICS

Benedict Anderson's *Imagined Communities*
Aristotle's *Politics*
Bernard Bailyn's *The Ideological Origins of the American Revolution*
Edmund Burke's *Reflections on the Revolution in France*
John C. Calhoun's *A Disquisition on Government*
Ha-Joon Chang's *Kicking Away the Ladder*
Hamid Dabashi's *Iran: A People Interrupted*
Hamid Dabashi's *Theology of Discontent: The Ideological Foundation of the Islamic Revolution in Iran*
Robert Dahl's *Democracy and its Critics*
Robert Dahl's *Who Governs?*
David Brion Davis's *The Problem of Slavery in the Age of Revolution*

Alexis De Tocqueville's *Democracy in America*
James Ferguson's *The Anti-Politics Machine*
Frank Dikotter's *Mao's Great Famine*
Sheila Fitzpatrick's *Everyday Stalinism*
Eric Foner's *Reconstruction: America's Unfinished Revolution, 1863-1877*
Milton Friedman's *Capitalism and Freedom*
Francis Fukuyama's *The End of History and the Last Man*
John Lewis Gaddis's *We Now Know: Rethinking Cold War History*
Ernest Gellner's *Nations and Nationalism*
David Graeber's *Debt: the First 5000 Years*
Antonio Gramsci's *The Prison Notebooks*
Alexander Hamilton, John Jay & James Madison's *The Federalist Papers*
Friedrich Hayek's *The Road to Serfdom*
Christopher Hill's *The World Turned Upside Down*
Thomas Hobbes's *Leviathan*
John A. Hobson's *Imperialism: A Study*
Samuel P. Huntington's *The Clash of Civilizations and the Remaking of World Order*
Tony Judt's *Postwar: A History of Europe Since 1945*
David C. Kang's *China Rising: Peace, Power and Order in East Asia*
Paul Kennedy's *The Rise and Fall of Great Powers*
Robert Keohane's *After Hegemony*
Martin Luther King Jr.'s *Why We Can't Wait*
Henry Kissinger's *World Order: Reflections on the Character of Nations and the Course of History*
John Locke's *Two Treatises of Government*
Niccolò Machiavelli's *The Prince*
Thomas Robert Malthus's *An Essay on the Principle of Population*
Mahmood Mamdani's *Citizen and Subject: Contemporary Africa And The Legacy Of Late Colonialism*
Karl Marx's *Capital*
John Stuart Mill's *On Liberty*
John Stuart Mill's *Utilitarianism*
Hans Morgenthau's *Politics Among Nations*
Thomas Paine's *Common Sense*
Thomas Paine's *Rights of Man*
Thomas Piketty's *Capital in the Twenty-First Century*
Robert D. Putman's *Bowling Alone*
John Rawls's *Theory of Justice*
Jean-Jacques Rousseau's *The Social Contract*
Theda Skocpol's *States and Social Revolutions*
Adam Smith's *The Wealth of Nations*
Sun Tzu's *The Art of War*
Henry David Thoreau's *Civil Disobedience*
Thucydides's *The History of the Peloponnesian War*
Kenneth Waltz's *Theory of International Politics*
Max Weber's *Politics as a Vocation*
Odd Arne Westad's *The Global Cold War: Third World Interventions And The Making Of Our Times*

POSTCOLONIAL STUDIES

Roland Barthes's *Mythologies*
Frantz Fanon's *Black Skin, White Masks*
Homi K. Bhabha's *The Location of Culture*
Gustavo Gutiérrez's *A Theology of Liberation*
Edward Said's *Orientalism*
Gayatri Chakravorty Spivak's *Can the Subaltern Speak?*

PSYCHOLOGY

Gordon Allport's *The Nature of Prejudice*
Alan Baddeley & Graham Hitch's *Aggression: A Social Learning Analysis*
Albert Bandura's *Aggression: A Social Learning Analysis*
Leon Festinger's *A Theory of Cognitive Dissonance*
Sigmund Freud's *The Interpretation of Dreams*
Betty Friedan's *The Feminine Mystique*
Michael R. Gottfredson & Travis Hirschi's *A General Theory of Crime*
Eric Hoffer's *The True Believer: Thoughts on the Nature of Mass Movements*
William James's *Principles of Psychology*
Elizabeth Loftus's *Eyewitness Testimony*
A. H. Maslow's *A Theory of Human Motivation*
Stanley Milgram's *Obedience to Authority*
Steven Pinker's *The Better Angels of Our Nature*
Oliver Sacks's *The Man Who Mistook His Wife For a Hat*
Richard Thaler & Cass Sunstein's *Nudge: Improving Decisions About Health, Wealth and Happiness*
Amos Tversky's *Judgment under Uncertainty: Heuristics and Biases*
Philip Zimbardo's *The Lucifer Effect*

SCIENCE

Rachel Carson's *Silent Spring*
William Cronon's *Nature's Metropolis: Chicago And The Great West*
Alfred W. Crosby's *The Columbian Exchange*
Charles Darwin's *On the Origin of Species*
Richard Dawkin's *The Selfish Gene*
Thomas Kuhn's *The Structure of Scientific Revolutions*
Geoffrey Parker's *Global Crisis: War, Climate Change and Catastrophe in the Seventeenth Century*
Mathis Wackernagel & William Rees's *Our Ecological Footprint*

SOCIOLOGY

Michelle Alexander's *The New Jim Crow: Mass Incarceration in the Age of Colorblindness*
Gordon Allport's *The Nature of Prejudice*
Albert Bandura's *Aggression: A Social Learning Analysis*
Hanna Batatu's *The Old Social Classes And The Revolutionary Movements Of Iraq*
Ha-Joon Chang's *Kicking Away the Ladder*
W. E. B. Du Bois's *The Souls of Black Folk*
Émile Durkheim's *On Suicide*
Frantz Fanon's *Black Skin, White Masks*
Frantz Fanon's *The Wretched of the Earth*
Eric Foner's *Reconstruction: America's Unfinished Revolution, 1863-1877*
Eugene Genovese's *Roll, Jordan, Roll: The World the Slaves Made*
Jack Goldstone's *Revolution and Rebellion in the Early Modern World*
Antonio Gramsci's *The Prison Notebooks*
Richard Herrnstein & Charles A Murray's *The Bell Curve: Intelligence and Class Structure in American Life*
Eric Hoffer's *The True Believer: Thoughts on the Nature of Mass Movements*
Jane Jacobs's *The Death and Life of Great American Cities*
Robert Lucas's *Why Doesn't Capital Flow from Rich to Poor Countries?*
Jay Macleod's *Ain't No Makin' It: Aspirations and Attainment in a Low Income Neighborhood*
Elaine May's *Homeward Bound: American Families in the Cold War Era*
Douglas McGregor's *The Human Side of Enterprise*
C. Wright Mills's *The Sociological Imagination*

Thomas Piketty's *Capital in the Twenty-First Century*
Robert D. Putman's *Bowling Alone*
David Riesman's *The Lonely Crowd: A Study of the Changing American Character*
Edward Said's *Orientalism*
Joan Wallach Scott's *Gender and the Politics of History*
Theda Skocpol's *States and Social Revolutions*
Max Weber's *The Protestant Ethic and the Spirit of Capitalism*

THEOLOGY

Augustine's *Confessions*
Benedict's *Rule of St Benedict*
Gustavo Gutiérrez's *A Theology of Liberation*
Carole Hillenbrand's *The Crusades: Islamic Perspectives*
David Hume's *Dialogues Concerning Natural Religion*
Immanuel Kant's *Religion within the Boundaries of Mere Reason*
Ernst Kantorowicz's *The King's Two Bodies: A Study in Medieval Political Theology*
Søren Kierkegaard's *The Sickness Unto Death*
C. S. Lewis's *The Abolition of Man*
Saba Mahmood's *The Politics of Piety: The Islamic Revival and the Feminist Subject*
Baruch Spinoza's *Ethics*
Keith Thomas's *Religion and the Decline of Magic*

COMING SOON

Chris Argyris's *The Individual and the Organisation*
Seyla Benhabib's *The Rights of Others*
Walter Benjamin's *The Work Of Art in the Age of Mechanical Reproduction*
John Berger's *Ways of Seeing*
Pierre Bourdieu's *Outline of a Theory of Practice*
Mary Douglas's *Purity and Danger*
Roland Dworkin's *Taking Rights Seriously*
James G. March's *Exploration and Exploitation in Organisational Learning*
Ikujiro Nonaka's *A Dynamic Theory of Organizational Knowledge Creation*
Griselda Pollock's *Vision and Difference*
Amartya Sen's *Inequality Re-Examined*
Susan Sontag's *On Photography*
Yasser Tabbaa's *The Transformation of Islamic Art*
Ludwig von Mises's *Theory of Money and Credit*

Macat Disciplines
Access the greatest ideas and thinkers across entire disciplines, including

AFRICANA STUDIES

Chinua Achebe's *An Image of Africa: Racism in Conrad's Heart of Darkness*

W. E. B. Du Bois's *The Souls of Black Folk*

Zora Neale Hurston's *Characteristics of Negro Expression*

Martin Luther King Jr.'s *Why We Can't Wait*

Toni Morrison's *Playing in the Dark: Whiteness in the American Literary Imagination*

Macat analyses are available from all good bookshops and libraries.

Access hundreds of analyses through one, multimedia tool.
Join free for one month **library.macat.com**

Macat Disciplines

Access the greatest ideas and thinkers across entire disciplines, including

FEMINISM, GENDER AND QUEER STUDIES

Simone De Beauvoir's
The Second Sex

Michel Foucault's
History of Sexuality

Betty Friedan's
The Feminine Mystique

Saba Mahmood's
*The Politics of Piety:
The Islamic Revival and
the Feminist Subject*

Joan Wallach Scott's
*Gender and the
Politics of History*

Mary Wollstonecraft's
*A Vindication of the
Rights of Woman*

Virginia Woolf's
A Room of One's Own

Judith Butler's
Gender Trouble

Macat analyses are available from all good bookshops and libraries.

Access hundreds of analyses through one, multimedia tool.
Join free for one month **library.macat.com**

Macat Disciplines

Access the greatest ideas and thinkers across entire disciplines, including

INEQUALITY

Ha-Joon Chang's, *Kicking Away the Ladder*

David Graeber's, *Debt: The First 5000 Years*

Robert E. Lucas's, *Why Doesn't Capital Flow from Rich To Poor Countries?*

Thomas Piketty's, *Capital in the Twenty-First Century*

Amartya Sen's, *Inequality Re-Examined*

Mahbub Ul Haq's, *Reflections on Human Development*

Macat Disciplines

Access the greatest ideas and thinkers across entire disciplines, including

CRIMINOLOGY

Michelle Alexander's
The New Jim Crow: Mass Incarceration in the Age of Colorblindness

Michael R. Gottfredson & Travis Hirschi's
A General Theory of Crime

Elizabeth Loftus's
Eyewitness Testimony

Richard Herrnstein & Charles A. Murray's
The Bell Curve: Intelligence and Class Structure in American Life

Jay Macleod's
Ain't No Makin' It: Aspirations and Attainment in a Low-Income Neighborhood

Philip Zimbardo's
The Lucifer Effect

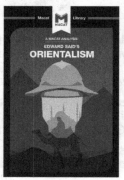

Macat Pairs

Analyse historical and modern issues from opposite sides of an argument. Pairs include:

HOW TO RUN AN ECONOMY

John Maynard Keynes's
The General Theory OF Employment, Interest and Money

Classical economics suggests that market economies are self-correcting in times of recession or depression, and tend toward full employment and output. But English economist John Maynard Keynes disagrees.

In his ground-breaking 1936 study *The General Theory*, Keynes argues that traditional economics has misunderstood the causes of unemployment. Employment is not determined by the price of labor; it is directly linked to demand. Keynes believes market economies are by nature unstable, and so require government intervention. Spurred on by the social catastrophe of the Great Depression of the 1930s, he sets out to revolutionize the way the world thinks

Milton Friedman's
The Role of Monetary Policy

Friedman's 1968 paper changed the course of economic theory. In just 17 pages, he demolished existing theory and outlined an effective alternate monetary policy designed to secure 'high employment, stable prices and rapid growth.'

Friedman demonstrated that monetary policy plays a vital role in broader economic stability and argued that economists got their monetary policy wrong in the 1950s and 1960s by misunderstanding the relationship between inflation and unemployment. Previous generations of economists had believed that governments could permanently decrease unemployment by permitting inflation—and vice versa. Friedman's most original contribution was to show that this supposed trade-off is an illusion that only works in the short term.

Macat analyses are available from all good bookshops and libraries.

Access hundreds of analyses through one, multimedia tool.
Join free for one month **library.macat.com**

Macat Disciplines

Access the greatest ideas and thinkers across entire disciplines, including

THE FUTURE OF DEMOCRACY

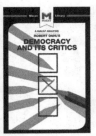

Robert A. Dahl's, *Democracy and Its Critics*
Robert A. Dahl's, *Who Governs?*
Alexis De Toqueville's, *Democracy in America*
Niccolò Machiavelli's, *The Prince*
John Stuart Mill's, *On Liberty*
Robert D. Putnam's, *Bowling Alone*
Jean-Jacques Rousseau's, *The Social Contract*
Henry David Thoreau's, *Civil Disobedience*

Macat Pairs

Analyse historical and modern issues from opposite sides of an argument.
Pairs include:

Macat Pairs

*Analyse historical and modern issues
from opposite sides of an argument.
Pairs include:*

INTERNATIONAL RELATIONS IN THE 21ˢᵀ CENTURY

Samuel P. Huntington's
The Clash of Civilisations

In his highly influential 1996 book, Huntington offers a vision of a post-Cold War world in which conflict takes place not between competing ideologies but between cultures. The worst clash, he argues, will be between the Islamic world and the West: the West's arrogance and belief that its culture is a "gift" to the world will come into conflict with Islam's obstinacy and concern that its culture is under attack from a morally decadent "other."

Clash inspired much debate between different political schools of thought. But its greatest impact came in helping define American foreign policy in the wake of the 2001 terrorist attacks in New York and Washington.

Francis Fukuyama's
The End of History and the Last Man

Published in 1992, *The End of History and the Last Man* argues that capitalist democracy is the final destination for all societies. Fukuyama believed democracy triumphed during the Cold War because it lacks the "fundamental contradictions" inherent in communism and satisfies our yearning for freedom and equality. Democracy therefore marks the endpoint in the evolution of ideology, and so the "end of history." There will still be "events," but no fundamental change in ideology.

Macat analyses are available from all good bookshops and libraries.

Access hundreds of analyses through one, multimedia tool.
Join free for one month **library.macat.com**

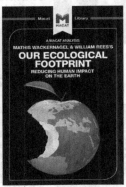

Macat Pairs

Analyse historical and modern issues from opposite sides of an argument. Pairs include:

ARE WE FUNDAMENTALLY GOOD - OR BAD?

Steven Pinker's
The Better Angels of Our Nature

Stephen Pinker's gloriously optimistic 2011 book argues that, despite humanity's biological tendency toward violence, we are, in fact, less violent today than ever before. To prove his case, Pinker lays out pages of detailed statistical evidence. For him, much of the credit for the decline goes to the eighteenth-century Enlightenment movement, whose ideas of liberty, tolerance, and respect for the value of human life filtered down through society and affected how people thought. That psychological change led to behavioral change—and overall we became more peaceful. Critics countered that humanity could never overcome the biological urge toward violence; others argued that Pinker's statistics were flawed.

Philip Zimbardo's
The Lucifer Effect

Some psychologists believe those who commit cruelty are innately evil. Zimbardo disagrees. In *The Lucifer Effect*, he argues that sometimes good people do evil things simply because of the situations they find themselves in, citing many historical examples to illustrate his point. Zimbardo details his 1971 Stanford prison experiment, where ordinary volunteers playing guards in a mock prison rapidly became abusive. But he also describes the tortures committed by US army personnel in Iraq's Abu Ghraib prison in 2003—and how he himself testified in defence of one of those guards. committed by US army personnel in Iraq's Abu Ghraib prison in 2003—and how he himself testified in defence of one of those guards.

Macat analyses are available from all good bookshops and libraries.

Access hundreds of analyses through one, multimedia tool.
Join free for one month **library.macat.com**

Macat Pairs

Analyse historical and modern issues from opposite sides of an argument. Pairs include:

Jean-Jacques Rousseau's
The Social Contract

Rousseau's famous work sets out the radical concept of the 'social contract': a give-and-take relationship between individual freedom and social order.

If people are free to do as they like, governed only by their own sense of justice, they are also vulnerable to chaos and violence. To avoid this, Rousseau proposes, they should agree to give up some freedom to benefit from the protection of social and political organization. But this deal is only just if societies are led by the collective needs and desires of the people, and able to control the private interests of individuals. For Rousseau, the only legitimate form of government is rule by the people.

Robert D. Putnam's
Bowling Alone

In *Bowling Alone*, Robert Putnam argues that Americans have become disconnected from one another and from the institutions of their common life, and investigates the consequences of this change.

Looking at a range of indicators, from membership in formal organizations to the number of invitations being extended to informal dinner parties, Putnam demonstrates that Americans are interacting less and creating less "social capital" – with potentially disastrous implications for their society.

It would be difficult to overstate the impact of *Bowling Alone*, one of the most frequently cited social science publications of the last half-century.

Printed in the United States
by Baker & Taylor Publisher Services